T0241886

Broken Pumps and Promises

Evan A. Thomas

Editor

# Broken Pumps and Promises

Incentivizing Impact in Environmental Health

 Springer

*Editor*
Evan A. Thomas
Department of Mechanical and Materials Engineering
Portland State University
Portland, OR, USA

.

ISBN 978-3-319-80396-8        ISBN 978-3-319-28643-3   (eBook)
DOI 10.1007/978-3-319-28643-3

Springer Cham Heidelberg New York Dordrecht London
© Springer International Publishing Switzerland 2016
Softcover reprint of the hardcover 1st edition 2016
Chapter 3 is published with kind permission of © 2016 by International Bank for Reconstruction and Development/The World Bank

Printed on acid-free paper

Springer International Publishing AG Switzerland is part of Springer Science+Business Media (www.springer.com)

# Foreword

In 2007, The Rockefeller Foundation convened a group of philanthropists, social impact leaders, and finance professionals at our Bellagio Center overlooking Lake Como to find new solutions on how to mobilize private capital to solve humanity's greatest challenges.

It might seem an odd place, seemingly so far away from the problems of the real world, to discuss such consequential topics. But that's precisely the Center's power: its serene location encourages those attending the conferences within its gates to dream big. And that's exactly what this distinguished group did, coining the term "impact investing" for a field that would help investors invest with the intention of both profit and social and environmental impact.

While the idea of using private capital to help solve humanity's greatest challenges wasn't itself novel, this new approach of double-bottom-line investing would lay the groundwork for new products and processes to channel more money, more effectively, towards these goals. And it comes at a critical time for philanthropy, as global philanthropic funds, even when combined with the development or aid budgets of governments, add up to billions of dollars. Meanwhile, the cost of solving the world's most critical problems runs into the trillions, including an estimated $2.5 trillion annual funding gap needed to achieve the Sustainable Development Goals (SDGs) in developing countries alone. Private capital is urgently needed in order to fill this gap and address pressing global challenges.

Since that meeting at Bellagio, the field of impact investing has taken root with the help of new infrastructure built with $40 million funded by The Rockefeller Foundation, including the creation of the Global Impact Investing Network, the rise of B-corporations, and the establishment of the Impact Reporting and Investment Standards and GIIRS analytics, now considered the "gold standard" for measuring a company or fund's social and environmental impact.

But there is still great opportunity for growing and developing the metrics and measurement tools that enable us to evaluate what is working and what is not. For those investors who seek to align payments with performance, innovations in both technologies and organizations will be needed.

At The Rockefeller Foundation, we are working to help support many of these innovations through Zero Gap, an effort dedicated to mobilizing large pools of private capital for social good. To do this, we are identifying the next generation of innovative finance products, partnerships, and processes that have the potential to create outsized impact. Employing a venture philanthropy model, Zero Gap supports early-stage design and leans heavily on collaboration and experimentation with both private and public sector partners. Whether it is pay for performance mechanisms or new institutional investment models, the solutions we are pursuing will all require objective data, feedback loops, and incentives for demonstrating that impact is actually achieved.

In the pages that follow, contributors discuss some of the emergent innovations in measuring the impacts of investment, with a specific look at poverty reduction. Edited by Professor Evan A. Thomas, this collection will be a valuable addition to the discourse on how we can better incentivize and evaluate impact across range of issues.

As an engineer and an entrepreneur working in global health, Professor Thomas has assembled compelling examples of technology, finance, and feedback that offer intriguing opportunities to close the gap between intent and impact. For example, the high adoption of mobile phones can help to accelerate the time it takes to make data actionable, while closing gaps in distance and subjectivity. Meanwhile, crediting systems, such as energy metering or carbon finance credits, can help align payments flowing from communities, donors, and investors with performance measures.

The development of such systems will be critical to supporting shared goals of mobilizing larger amounts of private capital to have more measurable and meaningful impact. Professor Thomas has edited much of this book while overlooking the same grounds as the pioneers of impact investing suggests that the Bellagio Center has once again inspired dreams that will transform lives around the world.

Judith Rodin, Ph.D.
The Rockefeller Foundation
New York, NY, USA

# Acknowledgments

This volume is a collaboration between all of the chapter co-authors as well as the numerous collaborators, funders and partners involved in the efforts presented. Particular thanks to Springer Editor Sherestha Saini and Portland State graduate student Emily Bedell. The Editor, Evan A. Thomas, dedicates this book to his wife, Lauren Alstot, and his mother, Anne Beirne.

# Contents

# Editor Biography

**Evan A. Thomas, Ph.D., PE, MPH** is an Assistant Professor and Director of the Sweet (Sustainable Water, Energy and Environmental Technologies) Laboratory, and a Faculty Fellow in the Institute for Sustainable Solutions at Portland State University. He works at the interface of engineering, environmental health, and social business, with professional experience working in government, industry, non-profits, and academia. He holds a Ph.D. in Aerospace Engineering Sciences from the University of Colorado at Boulder, is a registered Professional Engineer (P.E.) in Environmental Engineering in the State of Texas and holds a Masters in Public Health from the Oregon Health and Science University.

He is also a social business entrepreneur engaged in global health programs. Evan was a founding volunteer with Engineers Without Borders–USA in 2002, which led to co-founding Manna Energy Limited in 2007. In 2012, he co-founded SWEETSense Inc., an Oregon technology company. He also served as the Chief Operating Officer of DelAgua Health, a social enterprise partnered with the Government of Rwanda.

Prior to joining PSU, Evan worked as a civil servant at the NASA-Johnson Space Center in Houston, Texas. At NASA, he was a principal investigator and project manager in the Life Support and Habitability Systems Branch working on concepts and flight hardware for sustainable Moon and Mars spacecraft.

# Chapter 1
# Introduction

**Evan A. Thomas**

**Abstract** Global environmental health efforts are motivated by a sense of common responsibility and opportunity. These programs take forms large and small, from community groups to the World Bank. The methods likewise take varying, and sometimes competing forms, from watershed restoration to road building to community engagement, with funding provided by charities, loans, microfinance and big business. Once these projects are installed, typically the implementers are their own evaluators. When resources allow, some may invite external experts to visit the projects. Under the best of circumstances, funding is available to run a randomized controlled trial to rigorously evaluate if the projects are improving the intended environmental, health or other outcomes. But, usually sooner rather than later, the funding runs out for that particular project, and often organizations move on. This has resulted in sad statistics. For example, half of the water pumps installed in some African countries are broken a few years after they're installed. We propose an alternative – moving the mindset of funders toward pay-for-performance models of humanitarian and environmental interventions, backed by objective measurement tools and metrics. Instead of pushing money toward projects based on promises, pay interventions for successfully demonstrating impact that meets a stated intent.

**Keywords** Millennium development goals • Sustainable development goals • Impact • Intent • Pay for performance

## 1.1 The Intent to Impact Gap

The United Nations Sustainable Development Goals (SGDs) were announced with fanfare in September 2015. Replacing the retired Millennium Development Goals (MDGs), the 17 SDGs promise to deliver an ambitious range of impacts globally, including "End poverty in all its forms everywhere," "Ensure access to water and

E.A. Thomas (✉)
Department of Mechanical and Materials Engineering, Portland State University,
1930 SW 4th, Portland, OR 97201, USA
e-mail: evthomas@pdx.edu

© Springer International Publishing Switzerland 2016
E.A. Thomas (ed.), *Broken Pumps and Promises*,
DOI 10.1007/978-3-319-28643-3_1

1

sanitation for all," "Ensure access to affordable, reliable sustainable and modern energy for all", and "Revitalize the global partnership for sustainable development" (UN 2015).

While the intent is ambitious, what is less apparent is how impact and success will be measured. At release, the United Nations provided no objective standards or statistical indicators. These standards will no doubt be informed by the favorable interpretation of the progress made with the MDGs. In many cases, the United Nations claimed that the MDG goal targets were met. For example, the UN claimed to have, "met the target of halving the proportion of people without access to improved sources of water, five years ahead of schedule," (WHO/UNICEF 2012). Unfortunately, it has become apparent that the standards and measurements used for the MGDs were in many cases insufficient to actually meet these goals. As a result, the doubling-down with SDGs may equally fall short if measurement standards are not directly aligned with the impact intended.

Only a month after the SDGs were announced, the United States Government Accountability Office (GAO) released a report examining the United States Agency for International Development (USAID) efforts in water and sanitation. The title was straightforward – "USAID has Increased Strategic Focus but Should Improve Monitoring" (GAO 2015). The report commended USAID's water and sanitation efforts, but highlighted that, even by USAID's own metrics, they were likely overstating impact.

USAID's recommended standard and custom indicators include "Number of people gaining access to an improved drinking water source", and "Number of people gaining access to an improved sanitation facility". These indicators are intended to be collected annually for programs implemented in the previous year and have no meaningful consideration of monitoring over a period of years, measurement of water quality or sanitation level, or health impact. And yet, even with these demonstrably low quality indicators, USAID failed in many cases to collect data, and, in the view of the GAO, may have overstated their impact in claiming that millions have been provided access to safe water and sanitation.

Rather than an indictment of USAID or the United Nations, these examples instead highlight the status quo in delivering well-intentioned environmental health interventions. The finite and fickle flow of funds begets incentivizing new projects, and not sustained delivery of services.

## 1.2   Sustaining Impact

In contrast to piped water supplies, sanitation disposal or electrical grids in countries like the United States, service provisioning in many developing countries takes the form of household water filters, community hand-driven water pumps, improved wood, charcoal or kerosene cookstoves, and pit latrines. Access to these improved drinking water, sanitation systems and clean burning stoves could benefit the billions who suffer from diarrheal disease and pneumonia, two of the leading causes of death for children under five globally (UNICEF 2015). Billions of dollars are spent annually by governments, donors, non-profits and private sector institutions on technology interventions designed to provide these environmental services and address these public health issues.

The resilience of environmental service provisioning globally is dependent upon credible and continuous indicators of reliability, leveraged by funding agencies to incentivize performance among service providers. In the United States, these service providers are usually utilities providing access to clean water, safe sanitation, and reliable energy. However, in rural areas of developing countries, there remains a significant gap between the intent of service providers and the impacts measured over time.

This status-quo generally calls for finite funding and timelines of typically a few years to deploy, maintain and monitor such interventions. Impact is nominally evaluated by implementers directly. In some cases, funding may be available to employ health epidemiologists or development economists to run randomized controlled trials to rigorously evaluate if the projects are improving environmental, health or other outcomes. Yet, even when a positive impact is measured, the majority of these environmental service interventions are supported by implementers for only a few years. As a consequence, there is increasing evidence that much of the services provided in developing countries have failed to continue to positively deliver services.

Driving along a rural dirt road in many developing countries you see frequent evidence of this generous intent of global humanitarian aid agencies. Most tangible are hand driven water pumps that dot the landscape. These pumps are the concrete and steel outputs of a global intent to provide more clean water to more people. Thousands are installed every year, funded and implemented by organizations large and small. But, sadly, in many cases a flip of a coin may be your best judge of if the next water pump you pass will be surrounded by people, often women and children, filling their jerry cans, or if you'll see a decrepit artifact of wasted resource.

In rural sub-Saharan Africa, where hand pumps are a common technology, 10–67 % of improved water sources are non-functional at any one time, and many never get repaired (Foster 2013). While the proximate failures may be a leaky seal, a broken riser or a missing handle, these are only symptoms of the ultimate failure in how we fund, incentivize and monitor these efforts.

Presently, the impact of interventions may not always be aligned the intent originally sought. Improved regulations, standards and metrics that closely match intent, programs can be directly evaluated for compliance with those metrics and funders may incentivize and reward implementers for demonstrating impact.

Many organizations are now recognizing that a lack of objective data on program performance is contributing to a subsequent lack of accountability and misallocation of resources. Emergent tools and policy mechanisms may be able to respond to these issues. Improved and transparent feedback on the actual impact of global health and environmental programs may ensure the success of these efforts. Rather than infrequent data collection, more continuous feedback may improve community partnerships through continuous engagement and improved responsiveness. This approach seeks to raise the quality and accountability of these projects internationally by separating project success from advocacy. Additionally, by providing monitored data on the appropriateness and success of pilot programs, investors and the public can make more informed funding decisions.

In this book, we highlight some of the challenges in the current models of global environment and health efforts, and offer case studies that leverage feedback mechanism that can ultimately prove, and improve, impact. The status-quo is critically reviewed (Chap. 2) and evaluated by leading experts in development economics (Dennis Whittle of Feedback Labs, Chap. 4) and public health (Thomas Clasen of Emory University, Chap. 5).

On institutional levels, contributions from the Rockefeller Foundation, the Yunus Social Business and the World Bank provide frameworks for performance-based payments (Forward, Chaps. 3, and 16).

Programmatically, versions of these tools are demonstrated by the Freshwater Trust leveraging clean water crediting for ecological restoration (Chap. 7), and DelAgua Health using carbon credits to provide water and air quality public health interventions in Rwanda (Chap. 8 and 9).

Technologies such as cellular sensors and mobile money payments are use by Oxford University to deliver water pump services (Chap. 6), ethnographic researchers to evaluate sanitation interventions (Chap. 13), social enterprises including Sanergy Inc. to deliver sanitation services (Chaps. 12 and 14), and numerous small enterprises to deliver energy services (Chap. 15).

Finally, new models for monitoring, modeling and monetizing health impacts of interventions such as cookstoves are presented by Kirk Smith's research group at the University of California, along with program developers at CQuest Capital and NexLeaf Analytics (Chaps. 10 and 11). As Kurt Vonnegut said, "Another flaw in the human character is that everybody wants to build and nobody wants to do maintenance." With these innovations perhaps this flaw in global environmental health may soon be addressed.

# References

Foster T (2013) Predictors of sustainability for community-managed handpumps in sub-Saharan Africa: evidence from Liberia, Sierra Leone, and Uganda. Environ Sci Technol 47(21):12037–12046. doi:10.1021/es402086n

GAO (2015) Water and sanitation assistance: USAID has increased strategic focus but should improve monitoring. United States Government Accountability Office, Washington, DC

UN (2015) Sustainable development goals. 2015 time for global action for people and planet. United Nations, New York

UNICEF (2015) Committing to child survival: a promise renewed. United Nations Children's Fund, New York

WHO/UNICEF (2012) Millennium development goal drinking water target World Health Organization. United Nations Children's Fund, Geneva

# Chapter 2
# Performance Over Promises

Kristi Yuthas and Evan A. Thomas

**Abstract** Globally, stories of environmental health efforts are filled with good intentions and broken promises. Linking payments directly to long term social and environmental change can in some cases provide a solution. Pay for performance is now being used in a wide range of interventions and programs, but the potential of this approach is only beginning to be understood in the social sector. We explore theories that underlie pay for performance and lay the groundwork for understanding why and how this approach works. We then describe our Intent-to-Impact cycle—a four-stage model of Intent, Interventions, Evidence, and Pay for Performance that closes the loop between good intentions and impacts delivered. The challenge now is to use knowledge from this cycle to identify, explore, and learn from funding approaches that have and have not worked in important fields within the sector.

**Keywords** Pay for performance • Intent • Impact • Environmental health • Management theory • Global development

## 2.1 Pay for Performance

The common goal of nonprofits and social enterprises is to create positive social and environmental change. Yet the effectiveness of organizations in creating these changes varies greatly and the positive contributions of some organizations is debatable. In the absence of positive impact, some organizations are cost-ineffective in use of valuable resources that could be put to better use in making positive change.

When organizations fail to deliver promised impacts, donors can become skeptical and redirect their donations, taxpayers may push to reduce their governments' support of change efforts, and socially-oriented financiers may withdraw financial

K. Yuthas
School of Business Administration, Portland State University, 631 SW Harrison Street, Portland, OR 97201, USA

E.A. Thomas (✉)
Department of Mechanical and Materials Engineering, Portland State University, 1930 SW 4th, Portland, OR 97201, USA
e-mail: evthomas@pdx.edu

© Springer International Publishing Switzerland 2016
E.A. Thomas (ed.), *Broken Pumps and Promises*,
DOI 10.1007/978-3-319-28643-3_2

support or further emphasize financial returns. Such changes can greatly restrict the resources available to tackle devastating and persistent social problems.

### 2.1.1 Focus on Performance

One widely-promoted solution to this problem is increased accountability and transparency. Many funders are increasing their requirements for project monitoring and reporting and have encouraged more systematic evaluation and communication of activities and performance. These funders want evidence that their investees are doing the promised work and delivering the agreed upon outputs. But meeting these demands for accountability doesn't guarantee that the desired social changes have been achieved.

To ensure that these investments are having an impact, there has been an ongoing push toward providing hard evidence. The "gold standard" for reliable evidence comes from randomized control trials (RCTs). In this approach, measurements are taken before any action is taken, and groups are randomly assigned either to receive or not receive a funded intervention. At the end of the intervention, the organization or some external auditor measures whether the group that received the intervention is better off than the group that did not.

Although trials can be valuable, they are only useful for a fraction of investments, because they require very careful control of the intervention and they don't allow for any course corrections during the delivery of the program. This isn't possible in turbulent environments. And evidence about one intervention can rarely be generalized to another, because local conditions and populations vary so widely. This has left the sector with insufficient guidance on how to more efficiently and effectively address the needs of beneficiaries and to create the desired impacts.

Nonprofits and social enterprises in many fields are acutely aware of this challenge. As funds remain tight and problems remain massive in number and scope, the sector is looking for new ways to improve resource allocation and efficiency so that providers can do more with less. Pay for performance has begun to emerge in various forms in the social sector as organizations and their funders begin to recognize the potential for this paradigm.

At its core, pay for performance is the payment of money or other resources contingent on achievement of a performance goal. The increased recent interest in this approach results from the belief that funding can be designed to increase an organization's social performance through impacts such as improved quality of services, higher number of beneficiaries positively affected, or increased efficiency of service provision.

Donors have always cared about performance, and it has been common practice to link performance in one time period to funding in the next. However, like most ongoing funding, this performance-based funding has typically linked funding to inputs and activities rather than outcomes (Klingebiel 2012). This status quo is simplified in the following image.

Funders pay for successful performance in the delivery of services, such as the installation of solar panels or latrines. But whether these services have the intended

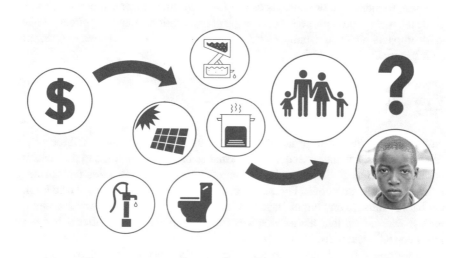

**Fig. 2.1** The status-quo in many environmental health interventions includes linear flow of funding that does not result in continuous or reliable feedback on impact for beneficiary communities

impact on the populations they're intended to serve remains unknown. So funders and service providers alike are left with little feedback for improving their operations or for more effectively directing resources (Fig. 2.1).

### 2.1.2   Elements of Pay for Performance

Pay for performance provides an approach and incentives to help ensure that the question of impact will be answered and the answer will be that positive performance outcomes have been achieved. Three key elements are important when designing and managing performance-based contracts:

Performance: The agreements made between partners will include process for measuring and evaluating performance. Outcome and/or impact goals are specified and related performance indicators are identified. This forces parties to be clear about both the end conditions they seek to achieve and the path through which these conditions flow from activities and outcomes.

Incentives: In performance-based contracts, at least part of the payment is linked to performance outcomes. Financial and non-financial incentives are developed to align the risks and objectives of the parties so that when an implementer produces the desired impacts, both the funder and implementer will benefit.

Risk: Linking rewards to performance creates increased risk for implementing partners. In traditional contracts, funders select implementers based on their past and expected future performance. The only recourse funders have for poor performance is the drastic measure of terminating the contract. In pay for performance, parties have incentive to refocus and innovate to continually improve performance outcomes.

These foundational dimensions of pay for performance arrangements affect and guide the intervention process, because the compensation received is directly linked to the impacts achieved. Contracts are carefully designed and executed to maximize their benefits.

### 2.1.3  Current Approaches

Pay for performance approaches can be characterized based on three broad categories: performance based aid, performance based incentives, and performance based contracting. Performance based aid refers to programs that link foreign assistance to program outcomes. For example, one program funded by the World Bank's Global Partnership on Output-Based Aid reserved final payments for water service until it was shown that the service was functioning six months after it had been installed (Klingebiel 2012).

Performance based incentives are programs that incentivize behaviors. In this approach, the funder has specific behavioral expectations for the beneficiary and/or service providers that are closely linked to desired outcomes. Typically these are individual behaviors, such as patient completion of health treatments.

Performance based contracting can be defined as a contractual agreement between a funder or purchaser of goods and services and the supplier of those services. A contract ties at least part of the payment to performance. Performance based contracting has become a standard feature of management in for-profit corporations. It includes a clear set of objectives and indicators, processes for gathering and evaluating performance data, and rewards or sanctions based on performance to which the contracting parties agree.

Pay for performance is used in a wide range of interventions and programs and the potential of this approach is only beginning to be understood in the social sector. The challenge now is to identify and explore funding approaches that have and have not worked in important fields within the sector. We begin by exploring the theories that underlie pay for performance to lay the groundwork for understanding why and how this approach works.

## 2.2  Theoretical Foundations

Linking pay to performance has a foundation in decades of research and experience with pay for performance, both within and between organizations. Early principal-agent economic models focused on the relationships between owners (principals) and workers (agents) under circumstances in which the two groups had different goals and the worker's efforts weren't always visible to the owners. Paying workers for their output, such as for the number of items they produced, was a way of aligning the incentives of the workers and owners and overcoming problems of unequal information and the potential for shirking.

In recent years, pay for performance models have increasingly been used to understand and manage the relationships between organizations and they have been widely studied in business-to-business and business-to-government settings. Performance based contracting (PBC) is used to align the interests of partners across the entire supply chain.

Pay for performance is perhaps especially important for social sector work because it stresses beneficiary (customer) value. By using financial tools to align interests, it ensures that all supply chain partners benefit when social impact goals are achieved. Contracting partners and beneficiaries have the incentive to work together to produce results, and as a result, have the incentive to share knowledge and work together to improve processes and pursue other innovations that can improve impact.

Principal-agent models use contracts to govern relationships. The contracts are usually designed to guarantee benefits for the principals or investors when outcomes are uncertain and they lack information. These contracts can specify what agents are supposed to do, or as we propose here, they can specify what agents are expected to achieve.

Management control theory, in part, combines both of these outcomes and focuses on coordination of activities across contracting partners. Processes are monitored throughout the project which enables greater control over the outcomes produced. Process monitoring requires a clear understanding of how inputs and activities result in outputs and desired outcomes. This approach requires the investor to have a better understanding of the value chain and greater participation in monitoring throughout an intervention.

Transaction cost economic theory can also shed light on the value of pay for performance contracts. This theory considers the contracting costs and the efficiency of maintaining partnerships across contracting periods. When financial rewards are linked to outcomes, implementers may be more willing to invest in assets and processes specific to one funder or intervention, because the mechanisms through which those investments will be recouped are more clearly specified.

Experience in for-profit businesses demonstrates that pay for performance contracts can not only provide greater benefits to investors, by generating the desired performance outcomes, but they can provide a host of other benefits as well. Shared goals and greater cooperation across the supply chain can lead to knowledge sharing, collaborative innovation, and greater performance returns on investments. These benefits can also be realized in participatory development initiatives if the interests of beneficiaries are embedded into contracts.

## 2.3  Aligning Intent with Impact

Achieving sustained social and environmental impact requires much more than good intentions. Parties working to create the change must develop a closed-loop system that ensures that the investments they make are monitored and managed

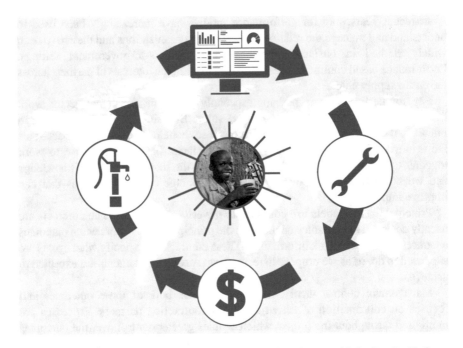

**Fig. 2.2** A closed-loop pay-for-performance model aligns direct impact with funding. In this figure depicting a water pump installation project, water point functionality is monitored, maintenance activities provided, and payments generated through evidence of service delivery

such that progress toward impact is visible, course corrections can be made, and achievement of impacts is rewarded. Here, we develop a model of how pay for performance can be used to turn good intentions into lasting social and environmental change. The model presented here includes four elements, as seen in the diagram below: intent, interventions, impact evidence, and pay for performance. The cycle then loops back to intent as the cooperating partners re-define and re-imagine their intentions and interventions to create greater impact in the next round (Fig. 2.2).

The intent-to-impact cycle has four elements:

- Intent—the desired social and environmental impact goals are identified and agreed upon.
- Interventions—the implementing partners plan and execute interventions to achieve performance goals.
- Impact Evidence—measurements of performance are gathered and analyzed.
- Payment for Performance—payments are made based on evidence of impact delivered.

This cycle provides a way for funders and implementers to close the loop by linking pay to the performance of those activities that provide the social and environmental benefits both parties seek. Linking pay to impacts has many advantages over traditional approaches in which implementers are paid based on their activities or on the delivery of products and services. Looking beyond these outputs to real, long-term impacts encourages partners to allocate resources to high-impact projects, to continuously innovate ways to achieve impacts more effectively and efficiently, and to make beneficiaries full partners in the co-production of impacts. This alignment is illustrated in the following figure.

### 2.3.1  Objectives of P4P

Pay for performance arrangements of all types are typically designed to achieve a similar set of objectives. First, they facilitate coordination between funders and providers to achieve end results for beneficiaries by designing incentives to ensure that goals among all three of these parties are aligned.

Second, by heightening the importance of performance outcomes and better defining the steps necessary for achieving them, pay for performance can help promote goal-oriented organizational systems. They can help create and support organizational systems in which achievement of goals isn't something attended to after completion of a project. Rather, it becomes embedded in the policies, processes, and even the culture of the organization. Third, pay-for-performance approaches promote external legitimacy. These arrangements signal that resources are being effectively managed, which can impact individuals from the parties involved but also potential future contracting partners. Because they demand accountability and transparency, at least to funders, these methods require systematization and discipline that can have numerous other organizational advantages.

Funders will typically seek to maximize the impact of any intervention and may have lofty or aspirational goals. Knowing that at least part of their payment will be determined by achievement of these goals, implementers will have the incentive to make sure that these goals are realistic, measurable and achievable.

### 2.3.2  Element 1: Intent

The first element, and the key to the pay-for-performance model is intent—the intended social and environmental changes that define and drive the model. In the first step in the intent-to-impact cycle, the collaborating partners establish the intent of their partnership. Before an intervention is planned, the partners need to agree on a clear set of output and impact goals. This is not the same as agreeing on expenditures, activities, or even goods or services to be delivered. For example, rather than agreeing that a certain number of water filters or malaria vaccines will be delivered, the parties will agree on the impact goals, such as disease reduction, that they want and expect to achieve. Each party must clearly define what a successful intervention would look like to them and eventually identify the kind of evidence that would convince them that long term success had been achieved.

This can lead to very different conversations than those focused on behavioral expectations. And the process can be quite enlightening, as it can reveal differences in intentions and perspectives among parties and even among members of one of the parties. For example, microfinance investors have very different intentions even as they agree on the means for achieving desired outcomes. If the focus is on short term operational goals, like granting a certain number of microloans or achieving a targeted portfolio value, the true intent of the loans may not be clear. But when intent is made visible, different perspectives will become apparent. Some investors seek to provide financial services in underserved markets. Others go further and seek poverty alleviation. Still others seek other outcomes such as women's empowerment, community cohesion, or improvements in environmental health. Clarifying the intent of an intervention can therefore help funders and operators to be more transparent about their objectives and formulate partnerships based on a better understanding of individual and mutual expectations and goals.

Voices of beneficiaries and other stakeholders affected by the proposed changes are also important in this process. The fact that interests and utility don't translate well across cultures has been well documented, and it is important to develop an understanding of whether the impact goals of funding and operating partners are consistent with the interests of their would-be beneficiaries. In one study, for example, it was shown that hungry people didn't maximize caloric intake when they had additional money to spend on food. Instead they spent part of the money on high quality food items they couldn't normally afford (Banerjee and Duflo 2012). In other studies, the poor have been shown to spend money on culturally important activities like weddings or funerals, despite having insufficient money to meet basic needs.

The advantage of using pay-for-performance approaches in such circumstances is that even if these preferences are not anticipated up front, these and other unintended consequences will quickly be identified as a result of performance monitoring, and if contracts are effectively designed, interventions can be re-designed to better achieve beneficial outcomes.

### 2.3.3   Element 2: Intervention

Intervention refers to the processes that will be enacted to achieve the intended long-term impacts. In traditional linear models of aid and development, the intervention comes first and impacts come second. The focus of many improved water initiatives, for example, is on how many water pumps will be installed and where they will be installed. If instead, impacts are prioritized, the focus might be on how many people will have improved health outcomes and how many cases of illness can be avoided.

Prioritizing impacts requires that each step in the intervention be designed with the goal of maximizing impacts. So a pump installation with a greater potential impact on health would be prioritized over one that reaches a greater number of people. When  impact is the principal concern, related organizational structures,

**Fig. 2.3** Logic model depicting the logical sequence leading up to intended impacts

systems, and processes can be designed to maximize potential for achieving intended goals. This emphasis should result in a logic model that tightly couples each element with ultimate impacts. Figure 2.3 shows a standard logic model.

A clear and well-designed logic model is essential for achieving desired impacts. Inputs include the monetary and human resources that will be devoted to achieving the desired impacts. Processes are the behaviors and actions that are to be performed. Outputs are the goods and services to be delivered by the implementers, such as immunizations delivered or cookstoves installed. Impacts are the social and environmental changes created by the interventions. They include core issues such as health, poverty, security, environmental health and resource depletion. Impacts are sometimes divided into intermediate outcomes such as changes in the behavior or attitudes of beneficiaries and longer term progress toward ultimate social goals.

Historically, the focus for the vast majority of funders and implementers has been on outputs. Under an output-oriented approach, funders specify deliverables in terms of outputs, and implementers carefully monitor and manage their achievements in delivering these outputs. Implementers are held accountable for outputs, and outputs are central to accountability and reporting activities. Far less effort is focused on the long-term impacts of these outputs and how these outputs will produce lasting social and environmental changes.

The intent-to-impact cycle requires organizations to clarify the linkages between outputs and impacts. Participants must understand why their investments and actions are expected to lead to change and how they expect this change to materialize. They must also estimate how much impact can be projected to result from the outputs to be delivered. Making this clear up front is critical because the sequence of steps in the logic model will be used to identify milestones toward achieving impacts. These milestones will be monitored and may be linked to intermediate payments in the pay-for-performance model.

## 2.3.4  Element 3: Impact Evidence

Gathering evidence through effective metrics and monitoring programs is the third essential element for ensuring that intended impacts are realized. When pay is linked to performance, effective systems are needed for gathering evidence regarding execution and outcomes for each step in the logic model leading to impact. And the impacts themselves must be very clearly defined and measured. This evidence is the only way to know whether an investment is making a social impact and to know the scale of that impact.

Developing effective measurement processes requires a systematic approach to measurement. Epstein and Yuthas (2014) describe the process in detail. After the intent and interventions have been established, partners must first agree on how the results of the measurement system will be used. Some of the metrics will be used as a basic for financial rewards, but measurements should also be used to control activities, to advance learning, and to drive actions to continuously improve expected impacts.

Epstein and Yuthas (2014) further argue that while a broad range of metrics may be gathered, a small number of metrics most closely linked to performance milestones should be actively monitored and analyzed. These metrics form the basis for communicating with funders and other stakeholders. They will ultimately be used as a means for managing performance and improving impacts.

Typical measurement systems, even in well-established social sector organizations, tend to focus on monitoring the quantity and quality of goods and services delivered to beneficiaries. Pay for performance agreements require development of measurement systems that are capable of either directly or indirectly measuring the social impacts achieved.

In the most advanced measurement systems, intermediate and final impact metrics are integral to the day-to-day management of the organization and are well understood by managers and key stakeholders. Further, they are used to drive resource allocations and to inform strategic decision making to ensure continuous improvement in the achievement of social impacts.

### 2.3.5 *Element 4: Pay for Performance*

The final element is linking payments and other rewards to desired social performance. Carefully designed financial incentives help to ensure that funders and implementers are in alignment on the impacts they seek to create. As a result of this goal alignment, the parties are more likely to share information and work together to execute plans. In addition, all parties are incentivized to innovate when conditions change or milestones are not met. Because incentives are linked to ultimate outcomes, implementers have the freedom to make course corrections as problems are identified or as new performance-enhancing opportunities arise.

Incentives are used to motivate providers to provide high quality goods and services and to ensure that these outputs are resulting in the expected social and environmental outcomes. Incentives can be monetary or non-monetary and they take the form of rewards or sanctions. In addition to cash and other monetary resources, non-monetary incentives such as a strengthened relationship with the funder or recognition by funder and peer communities for success are also strong incentives.

Typically, incentives take the form of monetary rewards linked to specific performance targets. Evidence of achievement of these targets is typically gathered using one or more metrics that seek to effectively represent the desired long-term outcomes. By their nature, metrics can never fully represent the broader phenomena they purport to measure, and they can be subject to misinterpretation or manipulation under the wrong circumstances. In addition, funders seeking to demonstrate rapid results may preferentially fund projects that are quickly and easily measured. Thus pay for performance methods work best when the real goals of the parties are aligned and not aligned because one party seeks a payoff.

Incentives must be carefully constructed to avoid focusing too much attention on only the outcomes specified. Other positive or negative activities or outcomes should also be anticipated and monitored. Funders will typically seek to maximize the impact of any intervention and may have lofty or aspirational goals. Knowing that at least part of their payment will be determined by achievement of these goals, implementers will have the incentive to make sure that these goals are realistic, measurable and achievable.

When effective pay-for-performance arrangements are designed, participants will be in alignment on desired impacts, the processes necessary for achieving them, and the role of the implementer in achieving these outcomes. When implementers achieve performance goals, they benefit in two ways: they gain financially as a result of well-designed investment arrangements, and they more effectively meet their social performance objectives—their raison d'etre. Funders also benefit. Pay-for-performance arrangements ensure that each individual investment is logically linked to the desired goals and that investment performance is more transparent and incremental progress is visible. The wealth of information and insights provided as a result helps funders allocate resource investments in a manner that aligns with their own impact goals, and helps to ensure that each individual investment will maximize expected social returns.

## 2.4  Moving Toward Pay for Performance

Many organizations and funders aren't interested in outcome-based incentives because they believe they aren't needed or even that they are counterproductive. Implementers who are living day-to-day with programs and seeing positive outcomes first hand believe that they are already making optimal decisions regarding allocation of resources and execution of programs, and that they need no further incentives to ensure that their work is successful. For them, spending money on any type of monitoring or evaluation of outcomes is viewed as a diversion of funds away from programs and benefits for recipients, and will result in reduced rather than improved impacts.

While it is true that these organizations may be functioning optimally, evidence suggests that there are opportunities for improving outcomes in virtually any endeavor and better information about outcomes can inform and facilitate those improvements. Beyond this, clear evidence about what works can inform others working in the sector and the good works of one organization can be leveraged to improve others and greatly expand impact. A large number of organizations do recognize these benefits and are very interested in obtaining clear performance information, but are prevented from doing so by financial and other barriers.

## 2.4.1   Overcoming Hurdles

One of the biggest hurdles faced by organizations seeking information about their impacts is the difficulty of measuring outcomes. Although some organizations have such faith in the value of their work they find measurement to be a waste of resources, many believe that there would be value in measuring performance but it is simply too ill-defined or unstructured to be measured. Substantial work is in progress in many fields to overcome such problems. In microfinance, for example, the social performance task force (SPTF) has established a body of metrics for measuring many facets of impact in the sector. There is still much work to be done, but progress in the area is rapid.

Cost is another considerable barrier for many organizations that want reliable evidence about the effectiveness of their work. As we've discussed earlier, there are many reasons organizations are unable to invest in systems that would help gather this evidence and provide the feedback necessary to guide and improve their performance. Top among them is preference among funders for investing in the programs themselves, rather than in the infrastructure needed to measure and manage information. In addition, even within program investments, many funders seek to fund new projects and innovations and are less interested in supporting the ongoing administration and management of investments that are already in place.

While these hurdles are significant, means of overcoming them are on the horizon. One significant development that has the potential to both improve the effectiveness of measurements and reduce cost barriers is the introduction of information and communication technology (ICT). As technologies become less expensive to develop, install, and maintain, the amount of information that can be made available to program operators increases dramatically. For example, distributed and remote sensing technologies, such as micro-weather stations that gather and relay information about weather conditions have been effectively implemented in a variety of regions. This information can be used to inform agricultural inputs or determine when insurance payments should be made for low crop yields. ICT-A refers to the use of ICT for accountability, and there are many recent examples of using ICT to improve transparency, and ultimately governance of both public and private services.

One important benefit of using these technologies is to provide greater access to information for contracting parties that are remote from the intervention or otherwise lack information about activities and performance. For example, recipients of health services in Karnataka India were provided with smart cards that could be updated when services were delivered by health workers (Bhatnagar 2014). Reports were generated automatically from the smart card data, which eliminated the time required and potential inaccuracies associated with records kept by health workers. Those in charge of managing health care were able to receive more timely, accurate, and complete information regarding successes and failures.

## 2.4.2 Pay for Performance in Practice

Moving from intent to impact by implementing pay-for-performance methods is a distant prospect for many organizations working in the social sector. They believe it is too costly, too time consuming, or too difficult and is beyond their grasp. But increasingly, there are success stories of organizations that have moved from promises to performance by linking pay to desired long-term impacts.

This book is full of detailed examples of organizations that have achieved success in closing the loop. They follow the principles of the intent-to-impact model laid out here and they are achieving impacts. Also included in this book are stories of failure to achieve impacts and discussions of how those failures could be prevented or remedied by effectively linking social and financial goals.

The massive resources devoted in development to achieving social goals, although well-intentioned, often fail to deliver on the promised impacts. Addressing this problem by linking rewards to impacts could be seen as a common sense approach to improving social performance or could be seen as a moral imperative for making faster inroads into addressing the devastating circumstances faced by half of the people on the planet. Either way, valuing performance over promises by creating financial links between intent and impacts can dramatically change the way investments are made and the world's most intractable problems are addressed.

## References

Banerjee A, Duflo E (2012) Poor economics: a radical rethinking of the way to fight global poverty. Public Affairs, New York

Bhatnagar S (2014) Public service delivery: role of information and communication technology in improving governance and development impact, vol 391, ABD economic working paper series. Asian Development Bank, Metro Manila

Epstein M, Yuthas K (2014) Measuring and Improving Social Impacts - A Guide for Nonprofits, Companies, and Impact Investors. Berrett Koehler

Klingebiel S (2012) Results-based aid: new aid approaches, limitations and the application to promote good governance. Discussion paper, German Development Institute, Bonn, Germany

# Chapter 3
# Trade-Offs and Risks in Results-Based Approaches

**Claire Chase and Aidan Coville**

**Abstract**  The growth in results-based approaches (RBA) to development financing is motivated by the underlying assumption that these approaches will incentivize implementing agencies to more closely align their actions and exert more focused effort on achieving the objectives of the program that have been agreed upon by the recipient and donor. At the same time RBAs strengthen accountability of both recipient and donor, by generating objective evidence that the agreed upon results have been achieved. But it should not necessarily be taken for granted that they are a more efficient approach to development just because payments are tied to results. If donor and government objectives are misaligned, RBA can help focus government actions towards a clear target that they may not otherwise focus on, but this comes with a shift in the risk burden towards the government agency, which is likely to come with an implicit risk premium.

**Keywords**  Results based financing • Risk transfers • Principal-agent model

## 3.1  Background

The earliest example of mainstreamed supply-focused RBA at the World Bank comes from the results-based financing for health (RBF 2015) which has contributed US$537 million to improve health services in 32 developing countries. Until recently, experience using RBAs in water and sanitation has been limited. A review undertaken by the World Bank in 2010 indicated that less than 5 % of its output-based-aid (OBA) portfolio was in water and sanitation (Mumssen et al. 2010). This

C. Chase (✉)
The World Bank, 1818 H St, NW, Washington, DC, USA
e-mail: cchase@worldbank.org

A. Coville
The World Bank, Abstrac1444 Fairmont St, NW, Apt 2, Washington, DC 20009, USA
e-mail: acoville@worldbank.org

© Springer International Publishing Switzerland 2016
E.A. Thomas (ed.), *Broken Pumps and Promises*,
DOI 10.1007/978-3-319-28643-3_3

has increased under the Global Partnership on Output Based Aid (GPOBA) (GPOBA 2015) –which includes 22 projects in water supply and sanitation. GPOBA is a US$190 million program that focuses on subsidizing service delivery to ensure it reaches the poor and marginalized, with funds flowing through direct service providers.

To further mainstream RBA into World Bank operations, the Program for Results Based Financing (PforR) was introduced in 2012 as a new lending instrument that links disbursement of funds directly to the achievement of specific program results, quantified as disbursement-linked indicators (DLIs). Together with funds from other sources, including other donors and development partners, PforR disbursements finance a borrower's expenditure program rather than individual transactions as under the existing Investment Lending and Development Policy Lending instruments. The instrument was designed to address the growing demand for sustainable results, and demand from client countries for program support. The program leverages a country's own institutions and processes to strengthen capacity and achieve results and is now the flagship results-based approach used by the World Bank. It currently has 27 operations totaling US$5 billion, of which a substantial proportion (17 %) of the funding falls under the Water Practice in the World Bank (see Fig. 3.1). Currently, there are three active operations in water supply, sanitation and hygiene in India, Mexico and Vietnam. Multilateral and bilateral agencies such as the Inter-American Development Bank and Department for International Development (DfID) have begun similar results-based approaches, such as DfID's Payment-by-Results (PbR) program.

The set of ongoing and potential RBAs to development financing is heterogeneous, cutting across multiple sectors, focusing on behavior change in service deliv-

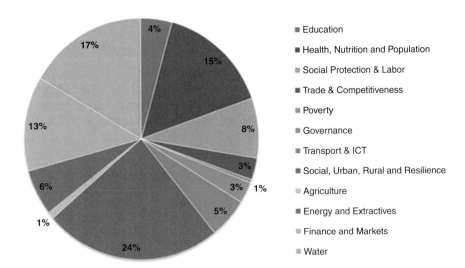

**Fig. 3.1** World Bank Program for Results portfolio and pipeline operations (World Bank 2014)

ery (e.g. Health RBF) or demand for services (conditional cash transfers), and setting contracts directly with service providers (e.g. GPOBA) or through national government ministries (e.g. PforR and PbR). They all, however, include a common thread – providing funding conditional on results achieved with the aim of improving the effectiveness and efficiency of programs to deliver results. By doing so, risk of poor performance is transferred from the donor to the recipient. The remainder of this chapter explores examples from the World Bank's PforR and DfID's PbR programs to highlight some of the less obvious sources of risk in RBA contracts, implications of this risk transfer on contract efficiency and the implicit tradeoffs to be made when deciding how to structure the agreement. We discuss the potential of technological innovation to reduce some of this risk through improved measurement and verification processes.

## 3.2 Using the Principal-Agent Model to Explain Risk Transfer and Uncertainty in RBA Contracts

The growth in results-based approaches (RBA) to development financing is motivated by the underlying assumption that these approaches will incentivize implementing agencies to more closely align their actions and exert more focused effort on achieving the objectives of the program that have been agreed upon by the recipient and donor. At the same time RBAs strengthen accountability of both recipient and donor, by generating objective evidence that the agreed upon results have been achieved.

Within the conceptual framework of a principal-agent model (Ross 1973) (Sappington 1991), the donor (principal) wishes to align the efforts of the recipient country government (agent) towards a specific output or outcome but is unable to directly observe the agent's actions that lead to this output. By linking payments to pre-specified, verifiable results, the donor and government objectives become more aligned, and the implementation risk associated with achieving the desirable outputs is transferred from the donor to the government.

In a simple model, which is useful for structuring the arguments later in this chapter, let's assume the donor is interested in output $y$ and the government generates this output through action $a$. The payment contract would then be $y = a + \varepsilon$ where $\varepsilon$ is a random error term including factors that are not under the government's control. Now the donor is able to set the RBA contract to $P = \mu + \beta y$ where $P$ is the payment to the government which is made up of an unconditional component ($\mu$) and a conditional payment that is a function of output $y$. $\beta$ can be set by the donor, where the higher it is, the more sensitive the final payment is to changes in the output generated.

One addition we can add to an otherwise standard principal-agent model for the purposes of this chapter is that, due to sampling error, we expect the true value of $y$ to be measured with noise (Holmstrom and Milgrom 1991). This means that $y = \hat{y} + v$ where $v$ is the random sampling (and potentially measurement) error induced through the verification process of an RBA project. Now the government faces the contract terms: $P = \mu + \beta \hat{y}$ which can be expressed in terms of the government's actions (what they can actually control) as $P = \mu + \beta a + \beta (v + \varepsilon)$. If we define government risk as increasing as the correlation between their controllable actions $a$ and payment $P$ decreases, then this formula shows that government risk increases when: (1) random factors that influence $y$ which are beyond the control of the government/contracting agency increase ($\varepsilon$); (2) the ratio of conditional to unconditional payments increases ($\mu$ vs. $\beta a$); (3) the rate at which actions generate payments increases ($\beta$); and (4) sampling and measurement error in results verification increases ($v$).

The higher the risk for the government to implement the program, the higher the risk premium would need to be for them to accept the terms of the agreement. Since the rationale for following an RBA rather than input-based approach to project financing is that it is more efficient to do so (measured by cost per unit of output achieved), and efficiency is a function of the associated risk premium, we should be interested in understanding what role the RBA risk transfer plays in practice and what opportunities there might be for reducing it. The rest of the chapter considers the practical manifestations of each of the points above. In particular we explore:

1. The challenges to RBA when the contracting partner (typically the Ministry of Finance in PforR projects) has limited control over generating outputs because these are under the control of local government (high $\varepsilon\varepsilon$));
2. The setting of conditional and unconditional payments to allow for practical service delivery ($\mu$ vs. $\beta a$); and
3. The practical implications of sampling error in verification procedures and what level of precision can be expected (what to realistically expect from $v$).

## 3.3 $\varepsilon$ Increases with Increased Levels of Accountability

The World Bank's PforR lending instrument and DfID's PbR program are designed to support government programs and institutional structures of the recipient. The rationale is that this allows the PforR to be used as a tool to strengthen capacity and systems within Government to deliver more effectively, inspiring an institutional focus on results, rather than trying to address specific local concerns in a piecemeal fashion. The implication is that, unlike GPOBA and Health RBF performance-linked incentives that directly incentivize the service provider, the contractual agreement is with the National Ministry of Finance (MoF) or equivalent. This means that there are now at least 3 levels of accountability and required changes in behavior. The donor incentivizes the MoF and central government who then sets up incentives for the local government to align around the objectives. In turn, local government leaders need to effectively incentivize employees to focus their attention on the agreed targets. Effectively this becomes a nested principal-agent setup with each level looking for effective approaches to change the behavior of the level below, but where each action taken by the respective agent to achieve their incentivized objective is subject to random errors outside of their control.

For example, DfID's payment-by-results (PbR) program with the Government of Tanzania provides an incentive to increase the coverage of functional rural water points across the country. The MoF is paid for each new and continually operating rural water point in the country. The MoF is then charged with transferring these funds to district (local government) offices across the country based on their performance. It is then up to the district offices to use their additional budget as they see best fit, which may include devising individual-level incentives for district water officials. The action of the water official has the most direct influence on the sustainability of a water point – by visiting the community and providing technical support they are expected to be able to reduce the likelihood that the water points within their purview break down. However, while this may be true on average, a multitude of external factors may influence how much their effort translates into actual improvements in the target. Local climate shocks, sub-national political instability, entry of new NGOs in the area, and randomly distributed hazard rates for engineering defects of the actual infrastructure can all be orthogonal to the efforts of the water official and enter the error term of the payment contract.

Going one step up, the district office aims to set the right environment and incentives for water officials to focus on maintenance issues. But even if they are able to set an optimal incentive scheme, they will be subject to random shocks such as staff turnover, disability or poor health of employees, or breakdown of district office transportation equipment, for instance. Finally, the MoF, while engaging districts through fiscal incentives, will have their effectiveness influenced by various country-wide shocks such as inflation, national economic shocks, or changes in political priorities across other sectors. In the setting described here, all of the risk associated with each level of government is included in $\varepsilon$ which the MoF is ultimately responsible for. Starting earlier in the system to address root challenges to implementing agent inefficiencies (contracting with the MoF) has the potential to generate system-wide changes to governance, but also comes with substantially increased risk to the contracting agency which ultimately translates into a larger risk premium, or cost per unit of output targeted.

## 3.4 Considerable Heterogeneity in the Setting of $\mu$ and $\beta\alpha$ Exists in Practice, Without Supporting Evidence to Rationalize the Ratio

A key motivation of the PforR is that Bank financing can be used as leverage to improve the efficiency of government programs. In practice, the proportion of conditional to unconditional financing of programs ($\mu$ vs. $\beta a$) varies considerably, and depends on various factors such as the number of development partners and the proportion of overall development assistance in a country. A high ratio of conditional ($\beta a$) to unconditional ($\mu$) financing increases the risk to government that actions that do not result in verified outputs will go uncompensated. Furthermore, outputs that require substantial upfront investment and take longer to achieve can cause a delay in payments to the program, resulting in insufficient levels of financing to keep a program running.

The ratio of conditional to unconditional financing of programs varies considerably. DfID's payment-by-results program to support Tanzania's rural water supply contributes approximately 54 % of DfID's £140 million funding basket, while the remainder is provided as input-based funding. Program support under the PforR for Rural Water Supply and Sanitation in the Red River Delta region of Vietnam was designed to contribute 77 % of the total budget envelope of the National Target Program for Rural Water Supply and Sanitation (NTP3) in the region. In practice, the ratio has been much higher due to low levels of counterpart financing, which has resulted in implementation bottlenecks. For instance, one of the program DLIs – working household piped water connections – can only be measured and verified once construction of the water scheme has completed and connections to households have been installed. Procurement and construction delays together with other unforeseen circumstances can hinder progress towards achieving the targets, and delay payment. This may create distortions when extra effort is expended to achieve the results or funds are borrowed from other sources as a stopgap measure.

At the other extreme, 7 % of India's US$22 billion Swachh Bharat Mission program is in the form of conditional financing through PforR. A high ratio of unconditional ($\mu$) to conditional ($\beta a$) program financing implies a greater risk to the donor since withheld payment may not substantially impact the operation of the program. The risk to the donor of non-performance in this case is not trivial as it represents a US$1.5 billion loan. Given the wide range of options for mixing conditional and unconditional payments for a given program, and the heterogeneity of these choices in practice, the empirical evidence on the optimal proportion of conditional to unconditional financing to enable more practical program delivery and balanced risk transfer is lacking.

### 3.5    Sampling Error ($v$) Can Be Reduced Through More Precise Verification, But This Comes at a Cost and Is Impractical Beyond a Certain Level of Precision

Up to now we have assumed that the donor is interested in a particular measurable output such as functioning water points or latrine coverage. In reality, these are a means to an end, with the ultimate development objectives, for instance of the World Bank, being reduced poverty and increased shared prosperity. In this case, the contract a donor would like to be able to make with a government agency would be payment on the delivery of various outcomes such as improved health and reduced poverty. Although health and poverty outcomes are relatively easily measured, the problem is that the agency's actions ($a$) will have far less correlation with the outcome of interest ($y$) like improved health than with a more directly influenced output like functioning water points. This creates two problems: (1) the error term increases further and the donor may end up rewarding governments that have luckily had positive external shocks and punishing unlucky ones rather than rewarding true performance and (2) there may be underlying trends in these outcomes that have nothing to do with government action that will change over time anyway. A way to solve these problems is to run a randomized control trial to net out any external factors and time trends and identify the causal impact of the government's actions. While this is certainly possible on a case-by-case basis, the cost, logistics and feasibility mean that this is not a broadly applicable solution. In this case performance contracts inevitably resort to more controllable output measures for targets. However, even the verification of outputs is subject to bias and will inevitably be measured with error.

In the PforR, measurable and verifiable results are specified as Disbursement-Linked Indicators (DLIs) and defined during program preparation. Each DLI carries with it a verification protocol, which specifies how the DLI will be verified and how disbursement will be made on the basis of verification. However, unless we take a census of the population, which would be prohibitively costly, all estimates of population statistics will include sampling (and often measurement) error. In other words, averages estimated from a sample of the population are just that: estimates; and these estimates include a margin of error. In general, verification protocols have skirted the issue, but could be more explicit about linking payments to results within a confidence band. It therefore becomes important to weigh the trade-off between the need to be confident in the estimate, and the cost of higher precision.

Typically, projects are carried out at sub-national level, and across multiple administrative units and it might not make sense to estimate a population average over the entire project area, especially when resources are allocated at a more granular level. Verification also needs to account for these levels of disaggregation which effectively means ensuring a certain degree of precision at each level of disaggregation. As a rule of thumb to estimate sample size, the sample required to be representative at an aggregate level can be multiplied by the number of units at which the estimates need to be representative. For example, if a sample of $M$ households is needed to estimate coverage of latrines at the district level, the required sample to estimate coverage for $n$ villages in the district would be $M*n$ households.[1]

Even in large samples, we still run into challenges measuring outputs accurately which adds further noise to estimates. This is especially an issue for outputs that are multi-dimensional, or that include subjective measurements, such as cleanliness. Measurement error can be reduced through adequate training of enumerators and adherence to strict quality control procedures during the verification process.

The implications of sample size on estimate precision can best be illustrated through an example. Imagine the RBA contract is based on increasing latrine cov-

---

[1] In reality, the exact sample size requirements will be a function of the variance in the output measurement at each level of aggregation, but the rule of thumb provides a close approximation for planning purposes.

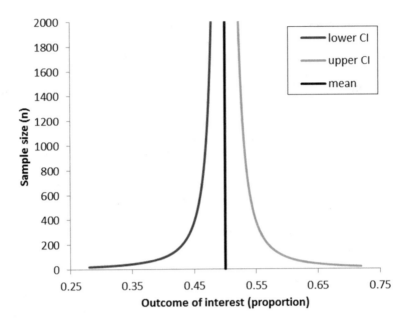

**Fig. 3.2** Sample size and precision

erage across the country. In the first case, the DLI is set at the district level – if a
district achieves 50 % improved latrine coverage across their district, they will be
eligible for a bonus payment. In reality, the only way we would know for sure if
the district in question achieved 50 % coverage would be to visit each household
in the district and see if they have an improved latrine. Since this is typically infea-
sible, a random sample is taken instead. Figure 3.2 below shows the target and the
95 % confidence interval associated with the estimate depending on the sample
size (*n*) used. Clearly at sample sizes below 100 the estimate is somewhat mean-
ingless – even if we found that the district sample showed 50 % coverage, we
know that the reality is that actual coverage may lie anywhere between 30 % and
70 %. The corollary to this is that we could measure a very high coverage in our
sample (e.g. 70 %) even if the true coverage lies below 50 % – a false positive – or
vice versa – a false negative. To reduce the likelihood of false positives or false
negatives, sample size needs to increase. As sample size increases, the confidence
band decreases, giving us more precise estimates. However, the rate of decrease
slows down as sample size increases, indicating that, at some point the cost of
additional sample doesn't warrant the very small increase in precision. Even at a
sample of 2000 we still have a margin of error of close to 2.2 percentage points on
either side of the mean.

Now let's assume we are willing to accept a margin of error of 5 percentage
points on either side of the mean which would require a sample size of 365 in our
example. If the DLI is set at the district level, and using the rule of thumb indicated

earlier, we would need a sample of 365 multiplied by the number of districts in the program. If, on the other hand, the DLI was set based on village-level coverage (each village that surpasses 50 % coverage receives a payment) then we need to multiply the 365 by the number of villages being considered in the program.

What if the margin of error is still too high? A slight increase in precision from 5 percentage points on either side of the mean to 4 percentage points on either side would require 50 % more sample, and further increasing precision to 3 percentage points on either side would require that the sample size almost triples. So, while there are large gains to be made in precision initially, the decreasing returns as $n$ increases means that we will inevitably have to live with some level of non-trivial sampling error. Further, while RBAs are often interested in attributing (and rewarding) good performance at a local level, verification costs may make this option intractable as the focus becomes more granular.

The question then becomes, how does one set payments in the presence of uncertainty derived from sampling error? One approach is to maintain that the estimate generated through verification will be interpreted as the best estimate of reality and payments will be made without regard to the uncertainty of the measure. This would implicitly transfer all random sampling error to the government and potentially increase their interest in ensuring a more precise verification process. Alternatively, payment could be made reflecting the confidence band, for instance by agreeing to provide payment as long as the estimate is not statistically less than the target. This would be a liberal approach to verification that would, on average, generate more false positives which would be a risk borne by the donor.

## 3.6 Improvements in Technology and Study Design Can Reduce Sampling and Measurement Error Without Necessarily Increasing Cost

Since data collection is costly but measurement and sampling error can undermine the integrity of a verification exercise, it is useful to have a set of tools that can help minimize the measurement and sampling error within a given budget. Technological advances and sampling designs beyond simple baseline/follow-up approaches offer the opportunity to reduce sampling error through increased frequency of observations and reduce measurement error through more accurate instruments.

Mckenzie (2012) illustrates that large reductions in sampling error can be achieved by increasing $t$ rather than $n$ in certain scenarios. In other words – focusing on multiple observations in time rather than across space. This is particularly useful when there is high variability over time for a given measure. For instance, if we are measuring an indicator of flow such as number of toilets built each month rather than a stock (proportion of households with an improved latrine) we may expect larger variations in time, in which case increasing $t$ may cost-effectively benefit precision of verification estimates. Collecting higher frequency data can be costly if it relies on standard enumerator-based approaches, but crowdsourcing information is becoming a popular approach to generating cheaper and more granular data as smartphones become more ubiquitous. For example, pilots in Tanzania and Liberia have begun testing a cell-phone application that gathers data from water users in rural areas to capture measures of infrastructure quality, use and functionality (McKenzie 2012).

Beyond increasing frequency, local sensors allow us to potentially increase the accuracy of indicators of interest, and in so doing, reduce measurement error. For instance, rather than relying on self-reporting of water use and functionality status for a particular water point, which can suffer from recall bias, water point sensors can send back real-time information of much more specific and accurate measures of water use. While the higher frequency and reduced measurement error that this provides makes these technological solutions an attractive option, there are still some barriers to scaling up given the high costs associated with sensor technologies currently, and the potential privacy issues if these are installed at the household level. The medium-term trajectory, however, is a positive one, where improved accuracy at lower cost associated with verification is within sight.

## 3.7   Conclusion

There has been rapid growth in RBAs over the past several years, and their appeal is evident. They make governments and donors more accountable that inputs translate into the agreed outputs, they help to focus efforts on tangible results, and as best practice can help strengthen government systems for monitoring, budgeting and planning. But it should not necessarily be taken for granted that they are a more efficient approach to development just because payments are tied to results. If donor and government objectives are misaligned, RBA can help focus government actions towards a clear target that they may not otherwise focus on, but this comes with a shift in the risk burden towards the government agency, which is likely to come with an implicit risk premium. If donor and government objectives are in sync in the first place, then these programs may have little impact on behavior change compared to the counterfactual, while still incurring a risk transfer and concomitant premium. As such, the relative efficiency of an RBA approach over input-based approaches as defined by cost per unit of output are likely to depend on the level of risk transferred to the government agency, which will vary across settings and contract structures. Value for money of RBA will depend on how willing donors are to pay this risk premium – and governments willing to accept it – to achieve the overall objectives of strengthened accountability, more robust government systems, and development impact. This chapter highlights some important sources of risk, how they may be perceived by both parties, and the tradeoffs that donors, governments and practitioners need to consider when setting payment structures and verification processes for results-based contracts in order to make these programs more effective.

# References

GPOBA (2015) Global partnership on output-based aid. World Bank. https://www.gpoba.org/

Holmstrom B, Milgrom P (1991) Multitask principal-agent analyses: incentive contracts, asset ownership, and job design. J Law Econ Organ 7:24–52

McKenzie D (2012) Beyond baseline and follow-up: the case for more T in experiments. J Dev Econ 99(2):210–221

Mumssen Y, Johannes L, Kumar G (2010) Output-based aid: lessons learned and best practices. World Bank, Washington, DC

RBF (2015) RBF health. http://www.rbfhealth.org

Ross S (1973) The economic theory of agency: the principal's problem. Am Econ Assoc 63(2):134–139

Sappington D (1991) Incentives in principal-agent relationships. J Econ Perspect 5(2):45–66

World Bank (2014) Program-for-results: an update. World Bank, Washington, DC

# Chapter 4
# How Feedback Loops Can Improve Aid and Governance

**Dennis B. Whittle**

**Abstract** If private markets can produce the iPhone, why can't aid organizations create and implement development initiatives that are equally innovative and sought after by people around the world? The key difference is feedback loops. Well functioning private markets excel at providing consumers with a constantly improving stream of high-quality products and services. Why? Because consumers give companies constant feedback on what they like and what they don't. Companies that listen to their consumers by modifying existing products and launching new ones have a chance of increasing their revenues and profits; companies that don't are at risk of going out of business. Is it possible to create analogous mechanisms that require aid organizations to listen to what regular citizens want – and then act on what they hear? Recently, many examples have emerged in which direct feedback from citizens has been solicited as input into both the selection and implementation of development initiatives. Not all have been successful, but some have led to significant improvements to outcomes. Being successful – having a *closed* feedback loop – requires not only that citizens be listened to, but that their voices be acted upon in the form of changes to aid programs. The chapter provides a set of principles that can be used by practitioners to design feedback loops that have a higher probability of success. It also provides a set of key conceptual issues that remain to be explored in depth by researchers, as well as a potential implementation roadmap for leaders of aid agencies.

**Keywords** Feedback loops • Accountability • Citizen voice • Global development

A version of this chapter was originally published by the Center for Global Development (http://www.cgdev.org/publication/how-feedback-loops-can-improve-aid-and-maybe-governance).

D.B. Whittle (✉)
Feedback Labs, 1913 12th St NW #C, Washington, DC 20009, USA
e-mail: dbw001@mac.com

## 4.1  Citizen Voice

"Ah, non – pas comme ca!" my new colleague Jean-Luc said sharply as he reached down and yanked out the rubber seedling. He held it up in front of the trembling farmer's face. "Bapak, tiga meter!" he growled. Three meters – that was the optimal spacing for planting rubber trees. Not the one and a half meters that he had just measured. Jean-Luc marched down the row of new plantings yanking out every other one, and I watched as the farmer's face grew more and more afraid. Finally, Jean-Luc said to me and the Indonesian official present "Let's go," and we got into the jeep and roared off down the road, back toward the capital of Jambi, a district on the Indonesian Island of Sumatra.

The date was late 1987, and this was my first "mission" to Indonesia. I had joined the World Bank only a year before, and spent my first 6 months helping negotiate a structural adjustment credit in Niger, where my main job was to tell government officials how to improve their water and sanitation systems. As I prepared for that earlier job, I was initially terrified and overwhelmed: I knew almost nothing about water and sanitation, and I had never been to Niger. Fortunately, I found that the Bank had hired a well-known engineering firm from France, and they had left behind numerous reports filled with all sorts of analyses. It turned out to be pretty straightforward just to read the reports, harvest the recommendations, and tell the government what to do in order to receive our low-interest structural adjustment credit worth tens of millions of dollars.

Having learned the ropes in Niger, I felt well prepared on that first trip to a farmer's field in Indonesia. I had read lots of reports from rubber experts, many of whom had spent the previous decades running plantations in places like Malaysia and Thailand and even Vietnam. It was clear that plantations could optimize the rubber yield per hectare by spacing the rubber trees at specific distances and applying fertilizer and pesticides at well-established rates at certain times of year. Building on this knowledge, in the 1970s the World Bank lent hundreds of millions of dollars to finance the planting of huge areas of Indonesia with tree crops, including rubber. The early projects had been implemented through state-owned plantation companies, which typically owned plots in the thousands of hectares. The Indonesian government was so pleased that it had asked the Bank to begin designing projects to help small farmers grow rubber, and the Bank responded enthusiastically, with several million more dollars of financing (The World Bank 1985).

These new smallholder projects recruited a lot of farmers who had never grown rubber trees. The objective was to get them to convert their fields of food and other crops primarily to rubber. They would be given a bank credit to cover their costs, and were expected to repay according to a certain schedule. Our models showed that, even after making their loan payments, farmers would have substantially higher incomes than before.

Many farmers, I later learned, were reluctant to be recruited, so they had to be "convinced" by local government officials, who needed to meet certain implementation targets. Others, like the farmer I had just met, embraced the opportunity to try something new that might increase his income. Yet, since he had no experience with rubber, he was wary of taking too much risk by converting all of his precious land. He had been told that he was required to plant 555 trees on his hectare of land, so he did a reasonable thing: he planted all of those trees on half of his land and he reserved the other half of his land for his usual food crops and to graze a few small livestock.

I was so outraged by what Jean-Luc did to the farmer that I complained about it to my boss, but he just shook his head and sighed. So I went back to my office in Jakarta, put my head down, and burrowed into my spreadsheets analyzing debt repayment schedules, future rubber yields by year, and the anticipated needs for rubber processing plants.

One day a few months later, I got a call from a young Englishman who was an advisor to the Indonesian Ministry of Planning. He said that he had just returned from a different province in Indonesia where the Bank was financing a similar project, though this one to plant coconut.

"You know, the way you are going about this is all wrong," he told me. "Farmers don't want to plant rubber and coconut so intensively because they need more of their land to plant crops that will give them immediate income rather than having to wait several years."

"But we give them credit to hold them over in those intervening years," I replied. "What's the problem?"

"Well many things," he told me, "including the fact the credit would not cover all their losses, and, even worse, local officials skim off a lot of that money. Plus, sometimes the farmers just like to plant certain things that are familiar to them and that they cannot easily buy in the market. Remember that they are not originally tree crops farmers, and they don't really understand whether the whole thing will work."

"Listen, why don't you come over to my office and I will show you the spreadsheets so you can see what's going on. I am surprised you weren't able to teach the farmer what to do," I told him.

"Well for heaven's sake – I'm not a tree crops expert," he replied. "I know very little about rubber and coconut planting."

"No wonder!" I exclaimed. "What are you anyway?"

"I'm an anthropologist by training, and I try to figure out about people's cultures and practices and how they like to live life."

I shook my head and did what every respectable World Bank economist at that time would do: I made excuses about not being able to meet him and hung up and continued with my spreadsheets and analysis. My colleagues and I later had vigorous and extended debates – including one very public showdown in a large meeting – with the British anthropologist and his fellow advisors about the best approach to smallholder tree crops and other agriculture sector issues. I was worried at first, because their arguments had a ring of truth, but eventually my colleagues and I prevailed because we held the power of the purse. We were able to lend the Indonesian government hundreds of millions more dollars, and our anthropologist friend was not.

## 4.2   Top Down to Bottom Up Aid

That encounter between Jean-Luc and the farmer haunted me for years. First and foremost, I could not shake the image of the farmer's face after his seedlings were ripped from the ground. He had been so proud when we arrived, and he looked so confused and scared as we pulled away in our jeep down the heavily rutted roads. And the battle we had with the government advisors had left a bad taste in my mouth. It had initially felt good to be on the winning side, but many of their arguments rang true, even at the time. My Bank colleagues and I had amassed enough sheer firepower to prevail, and I felt like a bully.

For the next many years, I continued to feel ill at ease with the work I was doing at the World Bank, where I participated in or led many projects and studies in several countries. I was keenly aware that I had the luxury of working with the largest aggregation of top development experts in the world. But the way we worked did not seem right. It took me more than a decade to begin to articulate the problem, and to understand how *poor information* and *perverse incentives* were the main causes.

The mental model in the development community when I began working in the field in 1984 was very top-down and expert-driven and went as follows: The developing world suffers from a severe shortage of both know-how and money. Official aid agencies such as the World Bank, the UN, and bilateral aid agencies need to aggregate the world's best expertise and money – and then deliver them to poorer countries. At that time, there was also a growing number of non-governmental organizations (NGOs). Though they get money from different sources (from donors rather than taxpayers), in practice many NGOs operate with a similar mindset: "We know what people need, we know how to deliver it, and we are here to give it to them."

There are two major flaws in this approach. First, we in the development community are not very good at knowing what people need or want. Second, we do not know how to deliver aid initiatives effectively on a consistent basis. Work by Bill Easterly and others has shown that the returns to trillions of dollars spent over the last six decades has had low returns (Easterly 2001, 2006). Even the more optimistic studies conclude that aid has had satisfactory returns only under limited circumstances.

The shortcomings of the top-down aid system reflect a broader problem with relying solely or primarily on experts. Philip Tetlock's celebrated recent work suggests that experts' predictions about the outcomes of complex situations and initiatives are poor – barely better than would be achieved by flipping a coin, and worse than would be achieved by applying various simple rules of thumb (Tetlock 2005). While most of Tetlock's work does not involve development experts, the outcomes experts in his studies were trying to predict were analogous to the predictions development experts make: If we do X, Y, and Z, then the outcomes A, B, and C will be realized.

During my time at the World Bank, most of my colleagues and I realized that many of our projects did not achieve the outcomes we had predicted. Unfortunately, we often had inadequate information to remedy the situation during implementation – not to mention the fact that we had strong incentives to avoid emphasizing or surfacing any shortcomings. The prevailing approach was to spend about a year or 18 months designing and appraising a project, then to spend 5 years implementing it, and finally to then evaluate the project. Although we did "supervise" projects during implementation, we could typically only spend a day or two at a sample of project sites. So we had to depend heavily on government implementing agencies for information.

For example, in the case of the rubber projects in Indonesia, the government agency not only reported back to us on farmers' attitudes toward the initiative ("the famers like planting rubber – it makes them better off!") but they also decided which sites we should visit. (And when we asked to see certain sites, they frequently demurred, saying that the road was washed out or that the site manager was at a training course). Only later, during the formal evaluation, did we discover that many tracts of land that we had financed had never been planted at all[10] – and that many unhappy farmers had failed to pay back their loans, saddling the government and banking system with big losses.

The problem was made worse by the institutional constraints we faced. World Bank budgets for project preparation, supervision, and evaluation were increasingly standardized across projects and across countries. As project managers, we faced clear incentives: namely, to get as many projects approved by the board as possible, and then to get them implemented in the allotted time frame, within the allotted budget, and without protests by the intended beneficiaries, public relations (PR) problems from local or international NGOs, or complaints from the government.

As long as we responded to these clear incentives, we got promoted and steady pay raises. But if we spent too much money monitoring the projects and slowed implementation down to try to correct for problems, our projects could be downgraded, and this would generally be reflected on our performance evaluations. If for some reason we were put in charge of a project with egregious and un-concealable problems, the rational thing to do was to ask for a transfer to a different project or even to a different country. Otherwise, we might be forced to negotiate with the government to formally redesign or even close the project prematurely, which was messy and time-consuming and (most damaging) would reduce the number of new projects we could prepare and send to the board (Phillips 2009). Further, the incentives were highly asymmetrical; there was no upside to making a project go better than anticipated (Pritchett 2002). If problems surfaced later, during the formal evaluations, it was not a huge problem because project managers had often moved on to a different country (Ostrom 2002).

The enemy of smooth project implementation was often citizen voice. It turned out that the poor rubber farmer I described above had relatively minor complaints compared to some others affected by aid projects in Indonesia. Far more serious were the complaints that thousands of villagers had about being forcibly relocated to make way for a dam that was being built. Some villagers were forcibly "transmigrated" to outlying islands with no physical or social infrastructure to in theory create a better life for themselves. A large number of these people were very unhappy about the lack of roads and functioning markets, schools, and health clinics.

In each case, the Bank assigned seasoned veterans to work together with the government to keep the situation from getting out of hand and to keep projects on track, with some refinements but rarely major modifications. The goal was, in the words of my colleague who had been assigned to manage the dam project, "To keep people's complaints down to a dull roar, and to avoid big stories, especially in the local newspapers, that might rile people up further."

Over the ensuing years, I watched a gradual change in the type of people hired by the Bank and other aid agencies. Increasingly I met new staff who had spent lots of time in the field and came to the agencies with a genuine understanding of and concern for real people. They came into their jobs with great enthusiasm, and whenever I met them it gave me glimmers of hope.

Alas, with few exceptions (Whittle 2006), these promising new staff soon learned that it was impossible to maintain enthusiasm and commitment in the face of the nearly overwhelming effort required to process projects through the Bank system. Ironically, the bureaucratic load typical of the 1980s was ratcheted sharply upward in the 1990s by a number of well-intended "safeguards" related to environmental, social, gender, and even procurement matters. These safeguards generated numerous requirements and led to even more studies, paperwork, and internal clearances – reducing even further the time that staff were able to spend in the field talking to real people.

The tragedy of this system is that most aid workers (certainly most of my colleagues at the World Bank) start out wanting to improve peoples' lives. They do the best they can, relying on the best information they can gather, to make a positive difference. Yet, over time they get ground down psychologically (and even physically) by the organizational constraints and incentives they face, and after a few years many of them lose touch with why they started this line of work in the first place.

Just before I left the World Bank in late 2000, I did an informal poll of many of my colleagues there, asking them "What proportion of your energy do you feel you are able to use in service of actually helping make the world a better place?" Their answers clustered tightly around 25 %. A typical response was "You know, I have not even thought about that question for so long. Most of my life and attention here are taken up by the need to write reports, attend meetings, get clearances, and book travel. I rarely get a chance to pause and ask whether it all makes sense, or whether I am making a real difference."

## 4.3   Impact Evidence

So how can we move forward beyond this outdated and ineffective mental model for development aid? There are no silver bullets, but some key principles are coming into focus.

In the last few years, there has been great excitement about the use of randomized controlled trials (RCTs) in development (Banerjee and Duflo 2012). RCTs were hailed as a way of overcoming an "inadequate understanding of poverty" and as "radical re-thinking of the way we fight poverty" (Banerjee and Duflo 2012). The idea is that by adopting the same standards of rigor applied in science, we would be able to find out "what works" and apply it around the world.

While the RCT movement is probably helping improve the rigor of ex-post evaluations of aid projects, its value in helping improve development outcomes on a wide scale is only modest. Despite the claims that RCTs drive medical advances, recent studies have questioned whether this is the case, for two reasons. First, medical companies have had exceptional difficulties replicating (in standard laboratory conditions) the results of pre-clinical RCT-based trials for things like cancer drugs. Even under laboratory conditions, scientists at the drug companies Amgen and Bayer, for example, were able to reproduce the results of only 11 % and 21 %, respectively, of the RCT-based trials they studied (Prinz et al. 2011; Begley and Ellis 2012).

Second, the drugs are administered under varying clinical conditions and to patients whose body chemistry differs significantly. According to a paper by Margaret Eppstein and colleagues, though "[m]any consider multicenter randomized controlled trials to be the gold standard of evidence-based medicine…results are often inconclusive or may not be generally applicable due to differences in the contexts within which care is provided" (Eppstein et al. 2012).

These drawbacks to RCTs in medicine echo the criticisms leveled at RCTs for development. Angus Deaton provided an early technical analysis of how the results from RCTs were unlikely to be transferrable to different contexts (Deaton 2010). Anyone who has managed aid projects realizes that there is a huge number of design and implementation parameters – and that it is maddeningly difficult to know which of these makes the difference between success and failure. In the preparation phase, we tend to give a lot of weight to the salience of certain factors, such as eligibility criteria, prices, technical features, etc. But during implementation, we realize that a thousand different factors affect outcomes – the personality of the project director, internal dynamics within the project team, political changes in the local administration, how well the project is explained to local people, and even bad weather can have major effects.

Development initiatives are not necessarily *complicated* but they are *complex* (Gribbin 2004). Complicated systems have many parts and aspects, but the outcomes can be predicted accurately if the initial conditions are known, since the parts themselves interact in a consistent and often linear way. Building a bridge over a wide river is complicated, as is building an airplane and most other engineering challenges, but experts are able every day to build bridges and airplanes that work reliably. By contrast, *complex* systems such as health care reform are very often completely unpredictable, and "infinitesimal causes can have enormous consequences" – a phenomenon known more popularly as the "butterfly effect (Horgan 2012). These complex systems – often involving human behaviors and interactions – are notoriously difficult to predict, much less control.

Lant Pritchett and Justin Sandefur have extended Deaton's analysis to show how RCTs can be expected to have little 'external validity' in development projects (i.e., the results are not transferrable beyond the initial context of the study) (Pritchett 2011; Pritchett and Sandefur 2013). Along with his colleagues Samil Samjih and Jeffrey Hammer, Pritchett goes on to propose a new way of rapid and ongoing iteration of project design during implementation (Pritchett et al. 2012). The idea is to start with a design considered reasonable, but focus a much greater proportion of available resources on finding out how well things are working in practice and then refining the design as you go.

Michael Woolcock argues that the more complex the initiative, the less likely RCT results are to be applicable across different contexts. The elusive idea of "best practices" is only valid in projects with "low causal density" – generally those whose outcome does not depend heavily on human behavior. He argues that using case study methodologies is critical to finding out what works in different situation. He suggests an approach similar to that which Eppstein et al. (2012) endorse for better medical research, namely "making it up as you go along: you work with others and learn from collective experience to iterate your way to a customized best fit" (Woolcock 2013).

It is true that RCTs can provide rigorous evidence of the impact of certain interventions under certain circumstances. But, in the end, randomized trials don't provide the tools for us to "radically rethink the way we fight poverty." Excessive faith in RCTs is in fact likely to reinforce the same top-down approaches that have had such poor results so far. They lend themselves heavily to initiatives where experts determined the desired outcomes ex-ante, marginalizing the voices of the people they are supposed to help. Instead of giving an incentive to local people to work together to forge solutions, they risk creating global "best practices" that ignore local knowledge. The tendency toward cookie-cutter prescriptions inhibits the emergence of local networks of problem solvers that experiment repeatedly until they find approaches that work.

## 4.4  Feedback Loops

Fortunately, there are new models that emphasize participation, accountability and feedback. Though these models are still in their formative stages, what many of them have in common is an assumption of rich and timely feedback loops that allow – or even require – implementing agencies to iterate constantly. Daron Acemoglu and James A. Robinson argue in their recent book *Why Nations Fail* that developing countries remain poor because their institutions are exclusionary and do not reflect the voices and needs of much of society (Acemoglu and Robinson 2012). They argue that as long as elites in these countries maintain near-monopoly control over public institutions, there will be only slow improvements in the quality of life for most people. Commenting on this book, Owen Barder notes: "If we think of politics as an endogenous characteristic of a complex system, then perhaps we have more hope of accelerating development by trying to tweak the internal feedback loops, and so shaping future system dynamics, than by offering exogenous solutions from the outside" (Barder 2013).

Acemoglu, Robinson, and Barder together suggest that helping citizens in developing countries have better voice through effective feedback loops would have high returns by increasing political inclusion and reducing the political monopoly of the elites. A key question is how aid projects can encourage the formation of these feedback loops without eliciting a backlash from the very elites whose power will be reduced by them.

The good news is that a number of experiments have been launched over the past few years to pilot new ways to collect and use feedback. The International Aid Transparency Initiative (IATI 2015) and Publish What You Fund (GCAT 2015) have made information on official aid available on a much wider scale. The same is true for non-governmental organizations, thanks to the BRIDGE project, launched in mid-2013 by GlobalGiving, Guidestar, The Foundation Center, and TechSoup Global. This latter project, partly funded by the Gates and Hewlett Foundations, is creating common data standards similar to IP addresses on the Internet that will enable people to compare data and information for about three million NGOs worldwide (Henderson 2013).

I recently reviewed a number of these experiments. Some were launched by non-profits, some by local governments, and others by official aid agencies. Some pilots failed completely, a few were very successful, and most had promise but lacked one or more of the elements required to either gather enough data or create pressures on implementing agencies to remedy the situation.

A small sample of these includes (FBL 2015):

- **World Vision (Uganda)** – Thirty school districts in Uganda used a Participatory Community Scorecard (PCS), developed in collaboration with World Vision. The PCS enabled communities themselves to develop the schools' performance criteria that the communities would monitor. For thirty other schools, experts defined the performance criteria that communities would monitor The schools for which the *community* defined performance criteria showed a .19 standard deviation increase in test scores, moving the average student from the 50th to 58th percentile in performance; increased pupil attendance 8–10 %; reduced teacher absenteeism by 13 %; and cost a total of $1.50 per student. The schools for which *experts* developed the criteria showed *no increase* in student test scores (Barr et al. 2012a, b).
- **Exposing Corrupt Politicians (Brazil)** – In this experiment, the results of audits of city finances were released to the public before elections. Compared to a control group without audits, Mayors were seven percentage points less likely to get re-elected when these audits showed corruption violations. The effect was more than doubled in towns that had radio stations that broadcast the audit results (PAL 2011).
- **Rapid SMS (Malawi)** – UNICEF trained local health workers to use an SMS-based tool that allowed them to report data on each child's health measurements.

This data previously took up to 3 months to be compiled on paper and sent to headquarters, and the data was mostly used for reporting and research purposes. Under the new system, the data took only 2 min to be entered into the phone and transmitted, and the platform responds immediately with tailored advice on nutritional needs for each specific child.

- **CheckMySchool.org (Philippines)** – This tool allows anyone – parents, students, teachers, administrators, or NGOs – to report problems at schools. These problems may range from absent teachers to missing textbooks to broken toilets. Comments and complaints can be channeled through email, SMS, Facebook, or a website and are viewable by the general public. The Department of Education has committed to taking quick action on complaints.
- **Crisis Response Map (Haiti)**: After the 2010 earthquake in Haiti, this online platform created by Ushahidi allowed anyone to provide updates or request help using SMS, email, phone, or Twitter. Thousands of reports were submitted, allowing rescue operations to reduce duplication of effort and focus on the areas most affected.
- **GlobalGiving Storytelling Project** – With support from the Rockefeller Foundation, GlobalGiving has collected stories from tens of thousands of people in Kenya about what they care most about. In one iteration involving four communities, GlobalGiving asked a panel of sixty-five aid experts and implementing agencies to guess the top priorities for the 2500 respondents. The expert panel guessed only half of the top six priorities of the community; and only one of the sixty-five experts correctly guessed the single most pressing issue (social relations) (Whittle 2010).

The number of effective feedback loop experiments in aid is still small, and though there are a priori reasons to expect them to improve aid projects, there have been few rigorous statistical evaluations of how well they are working. And, especially given our caution above about generalizing the results of RCTs, it would be foolhardy to try to draw any conclusions or "best practices." It may be more useful, at least at this stage, to identify some general principles and questions that are likely to inform the design of feedback loops that results in better outcomes.

Our review suggests that asking the following questions during the design stage is likely to increase the chance that the feedback loop will actually result in better outcomes:

1. What information is the feedback loop soliciting?
2. Who is most qualified to provide that information?
3. What incentives do those people have to provide the information? What are the costs and benefits that they perceive?
4. How will people provide that information? In person? Using certain technologies? Will the information be confidential or public?
5. Who are the intended recipients of the information and how will they get it?
6. What specific actions do we want the recipients of the information to take?
7. What incentives (carrot and/or stick) and capacity do the recipients have to take action? And how do we know action was taken?

As Pritchett and Woolcock argue above, the best way to approach these questions is probably just to start with a reasonable hypothesis, and then iterate based on experience. The answer to each question will depend critically on the context, for example the type of project, the sector, and even the type of implementing agency. A few studies that suggest reasonable places to start are starting to emerge.

With respect to Question 1 above, perhaps the most neglected feedback loop information is simply what people themselves care about most (Narayan et al. 2000). As the GlobalGiving Storytelling Project illustrated, experts and implementing agencies are often out of touch with what people want. Further, as Ben Olken has shown, direct participation by people simply in *choosing* projects can have a huge effect on improving satisfaction with outcomes and political engagement. In some contexts, the impact of giving people a say in the *choice* of which projects are implemented can be even greater than the impact of allowing people to monitor implementation (Olken 2010).

The World Vision initiative in Uganda described above suggests further insights into how to answer Question 1. In particular, allowing communities to define the "scorecard" for what gets measured even within a pre-existing program (in this case primary education) might be critical to improving outcomes. By contrast, the study found that allowing experts to define the scorecard produced no improvement in outcomes.

With respect to Question 2 above, other work by Ben Olken suggests that community participation in monitoring may work best for projects that do not require technical knowledge. Communities may be good at monitoring whether a school is working well (by observing, for example, whether the teachers are present and whether their children seem to be engaged and learning); but they are less good at monitoring expenditures on road construction, for which top-down audits are better able to evaluate construction techniques and raw material costs (Olken 2007).

On the other hand, Gray-Molina's work on hospitals in Bolivia suggests that, for more complex delivery systems such as health care, communities may be better at monitoring certain services than rules-based audits or evaluations related to proxies such as competitive hiring and staff supervision practices (Di Tella and Savedoff 2001).

In the *Exposing Corrupt Politicians* example in Brazil cited above, expert auditors were needed to actually find the information about corruption; regular citizens would have had a hard time doing that on their own (Question 2). In this case, the auditors' incentives to provide the information (Question 3) was their normal salaries and operating budgets – nothing else was needed, although we might anticipate local politicians will try to exert pressure on them not to report the information in the future. But the feedback loop would not have worked if the information had not been provided to citizen voters using local radio stations as the most powerful dissemination mechanism (Questions 4 and 5). And it was clear what actions the recipients of the information were going to take (Question 6) – they voted against the mayors associated with corruption.

In our initial review of feedback experiments, Question 3 – incentives to provide the information – was often a point of failure. In general there is a cost to information providers, including their time and money (for example, the cost of sending an SMS), and sometimes risk of retribution. In Tanzania, where half the public water points don't work, the organization Daraja created an initiative called Maji Matone to allow citizens to report on their water points. Over 6 months, only 53 reports were received (compared to an initial target of 3000). Although a formal analysis has not been done, it appears that there were at least three problems. First, men control the mobile phones, while women collect the water. Second, women apparently felt that there was a risk of retribution from local water officials. And third, women felt that their reports were unlikely to generate any remedial action.

With respect to action by feedback recipients (Question 7), Tessa Bold and colleagues found that feedback loops worked very differently depending on whether implementation was being done by a government or non-governmental organization. They took an NGO-led initiative that had improved educational outcomes in India and Western Kenya by identifying lagging students and assigning contract teachers to tutor them (Bold et al. 2013). They then replicated the approach throughout all provinces of Kenya, and randomized whether the program was implemented by an NGO or the government. Test scores rose significantly in the NGO-implemented programs, but not at all in government-implemented programs. It was not clear whether this was due to incentives or capacity constraints, although one can imagine a combination of both factors at work.

There are some reasons to believe that the new online aid intermediation platforms may have greater incentives and ability than traditional aid agencies to promote accountability through feedback loops, but many of the seven questions raised above remain to be addressed with these new mechanisms as well (Kapur and Whittle 2010).

As noted, there is no single answer to any of the above questions; the answers will depend on the context and will result from repeated iteration starting from reasonable hypotheses. One objective of future research should be to help generate these starting hypotheses. To this end, the research could fruitfully address five broad issues:

1. **How do we provide incentives for broad-based feedback?** At a minimum, people need to be technically able to provide feedback; they need to be able to afford it; they need to feel that it will make a difference; and they need to feel that they will not suffer retribution.

2. **How do we know that feedback is representative of the entire population?** In many places there is differential access to cell phones and even in-person meetings. The phenomenon of *elite capture*, where powerful local interests exert heavy pressure on elections and other decisions, is equally prevalent. Yet, our work to date suggests that these problems are generally lessened by new forms of feedback loops being introduced, and that more can be done to ameliorated if not eliminate bias.

3. **How do we combine the wisdom of the crowds with the broad perspective and experience of experts?** Effective feedback loops promise to increase the power of regular citizens in decision making about aid projects that affect them.

But these local decisions can often be improved by hearing the perspectives and advice of experts, who often have deeper knowledge of specific topics, and how specific approaches have worked in different countries. How can we use new feedback loops to create better conversations between citizens and experts about what investments and services would be have the biggest impact on well-being?

4. **How do we ensure there are strong incentives for aid providers, governments, and implementing agencies to adopt and act on feedback mechanisms?** Despite the growing number of feedback pilots underway, they still affect only a miniscule percentage of all government initiatives and aid projects. And a large proportion of the pilots to date have not led to significant changes that improve outcomes. Given the inherent incentives against accountability by big aid agencies, what combination of carrots and sticks will it take to bring about widespread adoption of feedback loops?

5. **What is the relationship between effective feedback loops in aid and democratic governance?** My research started out as an inquiry into how feedback loops could make aid agencies more accountable. But is it possible or desirable to separate feedback in aid from feedback more broadly in governance? Can or will promoting more effective citizen feedback in specific aid or government programs lead to greater citizen voice more broadly?

## 4.5   Conclusion

Promoting strong and timely feedback loops is key to making aid, philanthropy, and government initiatives more effective. Even in top-down schemes, benevolent experts and government officials have an interest in knowing how well implementation is proceeding, so that they can make mid-course corrections instead of relying on costly (expert) evaluations that come too late. But more broadly, feedback loops can also help us re-balance the way that development programs are formulated and conducted. Though progress has been made in listening to the voices of regular citizens, it is reasonable to guess that development assistance is still 80 % determined by experts and only 20 % by citizens. Good feedback loops could reverse this ratio and put the bulk of the decision-making power in the hands of regular people.

In the future, the "default" will be that aid officials need to demonstrate: (a) why they believe regular citizens actually want each proposed project, and (b) how citizen voice will be used to ensure high-quality implementation. Of course, there may be exceptions, for example certain types of policy projects or public goods with free rider problems, but the burden will be shift to the aid official to make the case for why citizen voice should not play a major role.

What is the fastest way to bring about this future? As discussed above, there are inherent disincentives for aid agencies and aid workers to seek out and act on citizen feedback. How to overcome these disincentives is a topic for another paper. But one thing I have learned in my nearly three decades of aid experience is that immediate and sweeping mandates rarely work – and sometimes they even backfire, by creating new compliance burdens that reduce the time available staff to address the real issues.

The best approach in the near-term is to provide carrots rather than sticks, at both the inter- as well as the intra-institutional level. Boards of governors of the different aid agencies should increase the resources made available to agencies that demonstrate a commitment to effective feedback loops. And within institutions, senior management should significantly increase the resources dedicated to experimentation and research, using the principles and addressing the conceptual issues sketched out above. Once successful approaches are well established, boards of governors as well as senior management should phase in mandates while providing a virtuous cycle of additional resources and incentives for agencies and staff that implement effective feedback loops.

To hasten this transformation, think tanks and citizen groups can rally public support for agencies that listen to people, and shine the light on agencies that persist with the old mental model of "experts know best." The Center for Global Development's *Commitment to Development Index*, the World Bank's *Doing Business* survey, and Transparency International's *Corruption Perception Index*, and Publish What you Fund's *Aid Transparency Index* are all examples of what might be emulated for feedback loops.

# References

Acemoglu D, Robinson JA (2012) Why nations fail: the origins of power, prosperity, and poverty. Crown, New York

Banerjee A, Duflo E (2012) Poor economics: a radical rethinking of the way to fight global poverty. Public Affairs, New York

Barder O (2013) It's the politics, stupid. http://www.owen.org/blog/6752

Barr A, Bategeka L, Guloba M, Kasirye I, Mugisha F, Serneels P, Zeitlin A (2012a) Management and motivation in Ugandan primary schools: an impact evaluation report. Parnership for Economic Policy Nairobi, Nairobi

Barr A, Mugisha F, Serneels P, Zeitlin A (2012b) Information and collective action in community-based monitoring of schools: field and lab experimental evidence from Uganda. http://citeseerx.ist.psu.edu/viewdoc/download?doi=10.1.1.372.6834&rep=rep1&type=pdf

Begley C, Ellis L (2012) Drug development: raise standards for preclinical cancer research. Nature 483:531–533

Bold T, Kimenyi M, Mwabu G, Ng'ang'a A, Sandefur J (2013) Scaling up what works: experimental evidence on external validity in Kenyan education. Center for Global Development, Washington, DC

Deaton A (2010) Instruments, randomization, and learning about development. J Econ Lit 48(2):424–455

Di Tella R, Savedoff WD (2001) Chapter 2: does voice matter? Participation and controlling corruption in Bolivian Hospitals. In: Diagnosis corruption: fraud in Latin America's Public Hospitals. Inter-American Development Bank, Washington, DC

Easterly W (2001) The elusive quest for growth: economists' adventures and misadventures in the tropics. MIT, Cambridge

Easterly W (2006) The white man's burden: whey the west's efforts to aid the rest have done so much Ill and so little good. Penguin, New York

Eppstein MJ, Horbar JD, Buzas JS, Kauffman SA (2012) Searching the clinical fitness landscape. PLoS ONE 7(11). doi:10.1371/journal.pone.0049901

FBL (2015) Feed Back Labs. http://www.feedbacklabs.org

GCAT (2015) Publish what you fund. Publish what you fund: the global campaign for aid transparency. http://www.publishwhatyoufund.org/

Gribbin J (2004) Deep simplicity: bringing order to chaos and complexity. Random House, New York

Henderson E (2013) BRIDGE to somewhere: a conversation with GlobalGiving, GuideStar, the Foundation Center, and TechSoup Global. Markets for good: information to drive social impact http://www.marketsforgood.org/bridge-to-somewhere-a-conversation-with-global-giving-guidestar-the-foundation-center-and-techsoup-global/

Horgan J (2012) Can engineers and scientists ever master "complexity"? Scientific American, New York

IATI (2015) International AID transparency initiative. IATI. http://www.aidtransparency.net/

Kapur D, Whittle D (2010) Can the privatization of foreign aid enhance accountability. J Int Law Polit 42(1):1143–1180

Narayan D, Patel R, Schafft K, Rademacher A, Koch-Schulte S (2000) Voices of the poor: can anyone hear us? World Bank, New York

Olken BA (2007) Monitoring corruption: evidence from a field experiment in Indonesia. J Polit Econ 115:200–249

Olken BA (2010) Direct democracy and local public goods: evidence from a field experiment in Indonesia. Am Polit Sci Rev 104(2):243–267

Ostrom E (2002) Aid, incentives, and sustainability: an institutional analysis of development cooperation: summary report. Sida, Stockholm

PAL (2011) Exposing corrupt politicians. Poverty Action Lab. http://www.povertyactionlab.org/publication/exposing-corrupt-politicians

Phillips D (2009) Reforming the world bank: twenty years of trial – and error. Cambridge University Press, New York

Prinz F, Schlange T, Asadullah K (2011) Believe it or not: how much can we rely on published data on potential drug targets? Nat Rev Drug Discov 10(9):712

Pritchett L (2002) It pays to be ignorant: a simple political economy of rigorous program evaluation. J Policy Reform 5(4):251–269

Pritchett L (2011) Development as experimentation (and how experiments can play some role) (Manuscript).

Pritchett L, Sandefur J (2013) Context matters for size: why external validity claims & development practice don't mix. Center for Global Development.

Pritchett L, Samji S, Hammer J (2012) It's all about MeE: using structured experiential learning ('e') to crawl the design space, vol 104. UNU WIDER, Helsinki

Tetlock PE (2005) Expert political judgment: how good is it? How can we know? Princeton University Press, Princeton

The World Bank (1985) Indonesia: smallholder rubber development II project. The World Bank, Washington, DC

Whittle D (2006) When official aid works. Pulling for the Underdog. http://www.denniswhittle.com/2006/11/when-official-aid-works.html

Whittle D (2010) If you can flip a coin, can you be an expert? Pulling for the Underdog. http://www.denniswhittle.com/2010/09/if-you-can-flip-coin-can-you-be-expert.html

Woolcock M (2013) Using case studies to explore the external validity of 'complex' development interventions. World Bank, Washington, DC

# Chapter 5
# Intent to Impact – Diluted Safe Water Monitoring

**Thomas F. Clasen**

**Abstract** In early 2012, the World Health Organization (WHO) and the United Nations Children's Fund (UNICEF) made an important announcement: "The world has met the Millennium Development Goal (MDG) target of halving the proportion of people without sustainable access to safe drinking water, well in advance of the MDG 2015 deadline" (WHO/UNICEF, Millennium development goal drinking water target World Health Organization. United Nations Children's Fund, Geneva, 2012a). Major news organizations heralded the accomplishment. The editors of The Lancet used the occasion to draw attention to underachievement other MDG targets but still acknowledged the water announcement as "some good news to celebrate" (The Lancet, Lancet 379(9820):978. doi:10.1016/S0140-6736(12)60412-7, 2012). There was little celebrating, however, among many who work at the intersection of water and health. This is because the way progress is measured on the MDG water target—by counting those that have access to "improved water supplies"—does not fully address water quality, quantity and sustainable access—key components of the target that are fundamental to human health. Overly simplistic metrics used to monitor progress on important health and development goals can be misleading. Monitoring that relies on poor indicators can exaggerate progress. Even worse, however, inadequate assessments of environmental health interventions can undermine the proper allocation of scarce resources to where they can best advance the intended goals. This chapter describes how existing monitoring of water supplies in low-income settings fails to address the key health-based conditions that the MDG water targets sought to encourage. Its aim is to use the inadequacy of international water monitoring as a case study to demonstrate the critical need to carefully align targets with conditions that are most likely to advance the overall goals, and the potential consequences of failing to do so.

T.F. Clasen, JD, Ph.D. (✉)
Rollins School of Public Health, Emory University,
1518 Clifton Rd, Atlanta, GA 30322, USA
e-mail: tclasen@emory.edu

© Springer International Publishing Switzerland 2016
E.A. Thomas (ed.), *Broken Pumps and Promises*,
DOI 10.1007/978-3-319-28643-3_5

**Keywords** Millennium development goals • Safe water • Monitoring • Joint monitoring program

## 5.1   Background

Target 7c of the MDGs called for "reducing by half the portion of people without sustainable access to safe drinking water" (UN 2000). Unfortunately, neither those that drafted the target for consideration by the UN General Assembly, nor the members of the General Assembly who voted for it, provided clear guidance on precisely what they intended by the language of the target. Neither did they define the key terms "sustainable", "access" or "safe".

However, the language used in the target has a history, and that history provides insights into its intent. The 1977 Mar del Plata Declaration by the United Nations, which launched the Water and Sanitation Decade (1981–1990), asserted the universal right to "access to drinking water in quantities and of a quality equal to their basic needs" (UN 1977). A dozen years later, the Global Consultation on Safe Water and Sanitation for the 1990s adopted the New Delhi Statement, calling on nations "to provide, on a sustainable basis, access to safe water in sufficient quantities … for all" (UNDP 1990). Agenda 21, the UN action plan for sustainable development that emerged from the United Nations Conference on Environment and Development held in Rio de Janeiro in 1992, expressed its goal in terms of "safe water" and set a minimum quantity of 20L/person/day (WSSCC 2000). Finally, the Secretary-General's report to Millennium Summit urged the adoption of a target to reduce by half the portion of people who lack "sustainable access to adequate sources of affordable and safe water" (Annan 2000).

Following adoption of the MDGs, the UN Secretary-General and the administrator of the United Nations Development Programme commissioned the UN Millennium Project, an independent advisory body, to identify the best strategies for meeting the MDGs. In its report, the Millennium Project's Task Force for Water and Sanitation defines "safe drinking water" as "water that is safe to drink and available in sufficient quantities for hygienic purposes" (UN 2005). The Task Force also endeavoured to shed light on the terms "access" and "sustainable": "access to drinking water requires the existence of infrastructure in good working order." It also noted that sustainable access implies "a type of service that is secure, reliable, and available for use on demand by users on a long-term basis." In terms of sustainability, the Task Force also noted that the "technology and processes should not result in environmental damage or other negative consequences…such as exposing people to health risks or creating pollution or degradation of the environment or downstream."

From this prior history, a few conclusions can be drawn about the intent of the MDG water target. First, with respect to quality, the intent is absolutely clear: drinking water must be "safe". In opting for this criterion over "acceptable", "acceptable quality" or "quality equal to their basic needs", the MDG target establishes an unequivocal mandate that water be free of pathogens. By focusing on "drinking water" rather than water "sources", the target also implied that the water be safe at the point of use, not just at the point of distribution. Notably, it is also expressed in terms of the population, not their households; this implies that that it was intended to cover more than just the home. Second, like most of the previous statements, the MDG expressly includes the concept of "access". The MDG Task Force on Water and Sanitation interprets this to include not only time spent procuring water—the traditional measure of access—but also in terms of (i) affordability, (ii) reliability, and (iii) the environmental impact of the supply (UN 2005). However, access also implies quantity, as the inverse relationship between distance to water supplies and the amount of water used has been consistently shown for more than 40 years (White et al. 2002).

## 5.2  Water and Health

Each of these priorities—quality, quantity and sustainable access—has been shown by research to be fundamental to optimizing health and development.

Systematic reviews of dozens of field studies have shown that interventions to improve water quality are effective in preventing diarrhoeal diseases—a leading killer of children (Esrey et al. 1991; Fewtrell et al. 2005; Clasen et al. 2006; Waddington et al. 2009). Field studies have consistently shown that in the absence of safe storage, even water that is safe at the point of distribution is subject to frequent and extensive contamination during collection, transport, storage and use in the home (Wright et al. 2004). Perhaps for this reason, interventions that improve or maintain water quality at the household level have reported higher levels of effectiveness against diarrhoea than conventional interventions at the source (Esrey et al. 1991; Fewtrell et al. 2005; Clasen et al. 2006; Waddington et al. 2009). There is also some evidence that water quality is associated with improved nutritional status (Dangour et al. 2013).

Water quantity is also fundamental to health. Systematic reviews have shown that interventions to improve quantity and access are also effective against waterborne diseases such as diarrhoea (Esrey et al. 1991; Fewtrell et al. 2005; Waddington et al. 2009). Moreover, by increasing water for personal hygiene, such interventions can also be protective against respiratory infections, trachoma and skin infections (Esrey et al. 1991; Rabie and Curtis 2006; Ejemot et al. 2008). In one recent review that looked specifically at studies that reported on the quantity of water in the home, researchers found that increased water usage for personal hygiene was generally associated with improved trachoma outcomes, while increased water consumption was generally associated with improved GI infection, diarrheal disease, and growth outcomes (Stelmach and Clasen 2015).

Water access has long been shown to be associated with the amount of water actually used (Cairncross and Feachem 1993). Round trip times exceeding 30 min for collecting water is associated with a significant decrease in water usage. This translates directly into adverse health outcomes. A recent systematic review showed a significant increase in illness risk in people living farther away from their water source (Wang and Hunter 2010). Another recent analysis estimated that a 15 min decrease in one-way walk time to water sources is associated with a 41 % average relative reduction in diarrhoea prevalence, improved anthropometric indicators of child nutritional status and a 11 % relative reduction in under-five child mortality (Pickering and Davis 2012).

Finally, the MDG target addressed the issue of sustainability. Research has shown that many of the improvements made to water supplies fail due to lack of maintenance or spare parts (Foster 2013). This is particularly true in the case of boreholes and other wells that require pumps to lift the water to the surface.

## 5.3   The Disconnect in Water Quality Monitoring

While quality, quantity and sustainable access are fundamental to the MDG water, however, they are not directly assessed by methods for monitoring progress on the MDG water target. A recent review summarizes the evolution of international water monitoring approaches and explains some of its deficiencies (Bartram et al. 2014). The simple explanation for the disconnect between international monitoring and the MDG water target is that the current monitoring system predates and was not designed to assess MDG targets directly. However, this disconnect has revealed important deficiencies with current water monitoring methods. The underlying explanation for these deficiencies is that monitoring the key conditions of quality, quantity and access on a global or even national scale is difficult, costly and beyond the capacity of existing monitoring systems.

When the time came to actually monitor progress toward the MDG water target, the Inter-Agency and Expert Group on MDG Indicators (IAEG) decided to rely on an existing system of reporting on water and sanitation that was never designed to capture the core components of the MDG water target (UN 2003). The WHO/ UNICEF Joint Monitoring Committee on Water and Sanitation (JMP) was organized in the 1980s, among other things, for the purpose of monitoring sector progress toward internationally established goals on access to water supply and sanitation. Its data comes from national household-level surveys, including the Demographic and Health Survey and the Multiple Indicator Cluster Survey, as well as local census data. Using such surveys to collect the data, however, required the JMP to rely on indicators for water coverage that minimally-trained survey administrators could identify and count; it did not have the tools or budget to measure water quality directly, or clear methods for assessing quantity and access. As a result, the JMP reports not on quality, quantity and access, but uses a proxy for these that is based on the type of supply the householder reports as its primary source of drinking water (WHO/UNICEF 2012a). For this purpose, it counts the source as "improved" if it consists of piped water, public taps, boreholes, protected wells, protected springs, or rainwater; unimproved sources include any other supply, including vendor-provided water, bottled water (in most cases), tanker trucks, unprotected wells and springs or surface water.

However, there is consistent evidence that "improved" water supplies are not necessarily safe. In the most recent analysis based on a review of 391 studies found that over a quarter of samples from improved sources contained fecal contamination in 38 % of 191 studies (Bain et al. 2014). The study concluded that access to an "improved source" does not ensure water is free of fecal contamination and that international estimates "greatly overstate use of safe drinking-water and do not fully reflect disparities in access." It called for enhanced monitoring strategy would combine indicators of sanitary protection with measures of water quality.

The JMP has been clear about the shortcomings of relying on the binary improved/unimproved typology to capture essential aspects of "sustainable access to safe drinking water". Its own field studies in six countries to explore options for assessing water quality directly found that except for some centrally managed piped water supplies, the so-called "improved sources" were often microbiologically and chemically contaminated, and that the level of faecal contamination was significantly worse at the household level (WHO/UNICEF 2010). One UN announcement about meeting the MDG water target made this clear: "Water quality surveys showed that many improved drinking water sources such as piped supplies, boreholes and protected wells, do not conform to WHO guidelines. On average, half of all protected dug wells may be contaminated, along with a third of protected springs and boreholes" (UN 2011). Even the 2012 JMP report acknowledges "it is likely that the number of people using safe water supplies has been over-estimated."

Moreover, although the JMP collects data on time spent collecting water to provide information on access (and thus, indirectly, quantity), this is not factored into the improved/unimproved characterization of water supplies that is the sole basis for scoring toward the MDG water target. A closer review of the JMP report shows that only a quarter of householders have water that is "piped on premises" (WHO/UNICEF 2012b). The other three-quarters travel once or more daily to collect their water, travelling a mean time of approximately 30 min round trip—a distance that that makes it unlikely they are procuring enough water for personal hygiene much less other productive uses such as irrigating gardens that could be used to improve nutrition (White et al. 2002). Moreover, in 71 % of all households without water on the premises, women or girls are mainly responsible for water collection, taking time away from caring for children (also predominantly done by them) or attending school and thus undermining other important MDG targets. Except for those that have achieved water that is "piped on premises", the claim that the MDG water target has been met says little about the extent to which we have actually improved "sustainable access" (WHO/UNICEF 2012b).

It is not clear why the JMP, which is doing much to improve its methods for monitoring and assessing water and sanitation, allows its current system of improved/unimproved water sources to be treated as a proxy for "sustainable access to safe drinking water." When the UN imposed its MDG water monitoring on it, the JMP could have acknowledged that it does not currently have the tools for measuring the key elements of quality, quantity and access, and continued to work toward a more complete monitoring system. This would have left the UN without a clear means of monitoring progress on the water target, but it would have avoided another premature claim of "mission accomplished".

### 5.3.1   Potential Policy Implications

Allowing "improved water supplies" to serve as the proxy for the MDG target of "sustainable access to safe drinking water" has several potential policy implications that could adversely impact health and development.

First, it exaggerates and distorts actual progress. The more obvious reason for this is because the monitoring system does not address the quality, quantity and sustainable access criteria that were intended by the MDG water target. A more comprehensive estimate that includes water quality criteria found that 1.8 billion people lacked "sustainable access to safe drinking water" in 2010 compared to the JMP estimate of 784 million who lacked "improved water supplies" (Onda et al. 2013). This resulted in a 10 % shortfall in actually meeting the MDG target. There is also a more subtle distortion based on the fact that the benchmark for water is community-level access rather than household-level access. Measured at the household level, the global deficit is as great for water as it is for sanitation (Cumming et al. 2014).

Second, current monitoring fails to address water access in schools, healthcare facilities, workplaces and other places outside the home (Cronk et al. 2015). As a result, there are reduced incentives for implementing water interventions in these locations, even though the density of the populations they might serve and the risk of contaminated water that these populations must rely on suggest that interventions in these non-household settings could be even more critical for improving health (Sobel et al. 1998). The proposed post-2015 water targets would address schools and heathcare facilities but not workplaces, markets, transportation stations and other non-household locations where great numbers of people—many of whom may be carriers of waterborne pathogens—often assemble and exchange microbes.

Third, current water monitoring creates incentives for sub-optimal programs and implementation that could adversely impact health. There are significant differences in "improved water supplies" and the quality and quantity of water they provide for use at the household level (Bain et al. 2014). By focusing solely on increasing coverage using approaches that meet the definition of "improved supplies", the current approach creates incentives for implementing lower cost solutions that count toward the target, such as of protected wells, communal tap stands and other supplies, rather than household connections that are more likely to improve water quality and quantity but are more costly. The current indicator also fails to address the sustainability of the solution, another focus of the MDG target. Lower-cost approaches that are not implemented with proper maintenance or other efforts to ensure that the supplies are available continuously and year-round could encourage users to revert to surface and other secondary water sources that are of poorer quality (Foster 2013).

Fourth, current approaches to water monitoring distort decisions about where resources should be prioritized. Poor households in remote rural locations or in urban slums are the most likely to be at risk of unsafe water supplies and and access to healthcare if necessary to deal with the consequences of relying on such facilities. At the same time, these same household are the most costly to reach with basic water services. Current monitoring ignores these increased risks, and creates an incentive for addressing "low hanging fruit"—urban and peri-urban locations where the cost per household served is lowest and there is potential for cost recovery through user fees. Moreover, by using a single indicator that treats a wide variety of service levels as equivalent and does not address quality, quantity and sustainable access, the current approach yields little information about actual deficiencies in these key aspects of water supplies in existing populations. This limits the potential for policymakers to identify who could best benefit from improvements.

Finally, by treating "improved water supplies" as equivalent to "sustainable access to safe drinking water", the current monitoring system will allow many countries to claim that they have met the MDG water target. Given limited resources, policymakers in these settings may choose to divert resources away from continued investment in water and into other priorities.

## 5.4  Looking Forward

Over the years, considerable efforts have been undertaken to expand the scope of international water quality monitoring in order to address the key components of quality, quantity and sustainable access that are vital to improve health. The third edition of the WHO's Guidelines for Drinking Water Quality recommend a more comprehensive approach that addresses quality, coverage, quantity, continuity, and cost (WHO 1997). A health-based approach using water service levels was proposed in 2003, and a human rights based approach adopted in 2008 (Kayser et al. 2013). However, the fundamental dichotomous "improved/unimproved" approach is still largely employed by the JMP, with a new rung added to distinguish "piped" water supplies to the household.

There is increasing recognition of the need for a more comprehensive "service quality" or "service ladder" approach that accounts for the different levels of service provided by various drinking water and sanitation facilities, and their associated benefits (Bartram et al. 2014). Bartram and colleagues argue that at a minimum, this system should distinguish piped, household connections from other types of improved water supplies. They also recommend that water source functionality and reliability should be part of the analysis. Finally, for households without access to reliable household-level piped supplies, they recommend some measure of the safety of household drinking water storage methods, though it is not clear if this would constitute some type of water safety plan compliance or actual testing of water quality. It not clear whether this ladder would somehow incorporate measures of water quantity or actual use.

Perhaps the more important key to comprehensive water quality monitoring, however, is the indicators for the targets. A recent review has described how a large variety of indicators have been used to assess water source/technology type (including whether categorized as "improved," "unimproved," community source, or on-plot water), accessibility, water safety (quality and sanitary risk), water quantity, reliability or continuity, affordability, and equity (Kayser et al. 2013). While the review explored the potential for combining these indicators into a comprehensive framework, it concluded that the scientific basis for doing so was still lacking and that further research was necessary.

Nevertheless, expert groups assembled by the JMP have proposed a set water targets as part of "Sustainable Development Goals" (SDGs). For "basic drinking water" the proposed indicator uses the existing definition of "improved water supplies" combined with an accessibility criterion: "using an improved source with a total collection time of 30 min or less for a round trip including queuing." "Intermediate drinking water" includes an improved drinking water source on premises with discontinuity less than 2 days in the last 2 weeks; with less than 10 cfu E. coli/100 ml year round at source; accessible to all members of the household at the times they need it. Significantly, only the 2040 goal moves beyond current reliance on water service levels as a proxy for actual water quality testing. While these represent progress, they continue to fall short of more comprehensive water monitoring. There is also continued uncertainty about how these indicators will actually be measured.

## 5.5   Conclusion

Water is also fundamental to health and human development. When the UN Millennium Task Force on Water and Sanitation analyzed the contribution of water toward each of the MDGs—including eradicating extreme poverty and hunger (goal 1), reducing child mortality (4), and combating major diseases (6)—it concluded that "for many of the targets, it is difficult to imagine how significant progress can be made without first ensuring that poor households have a safe, reliable water supply and adequate sanitation facilities" (UN 2005). In order to meet these goals, however, it must be supplied in a manner that meets minimum requirements for quality, quantity and sustainable access that were envisioned by the MDG water target.

As management gurus are wont to advise, "what gets measured gets done". Current metrics for assessing water supplies fail to incorporate key indicators to ensure that they address these key conditions of quality, quantity and sustainable access. While the post-MDG targets represent progress by establishing additional indicators, more comprehensive monitoring is required. In the meantime, advocates for safe, reliable drinking water for all must hope that the premature celebrations about meeting the MDG water target do not slow progress in providing this basic human resource to all.

## References

Annan K (2000) We the peoples: the role of the United Nations in the 21st century. United Nations, Department of Public Information, New York

Bain R, Cronk R, Wright J, Yang H, Slaymaker T, Bartram J (2014) Fecal contamination of drinking-water in low- and middle-income countries: a systematic review and meta-analysis. PLoS Med 11(5):e1001644. doi:10.1371/journal.pmed.1001644

Bartram J, Brocklehurst C, Fisher MB, Luyendijk R, Hossain R, Wardlaw T, Gordon B (2014) Global monitoring of water supply and sanitation: history, methods and future challenges. Int J Environ Res Public Health 11(8):8137–8165. doi:10.3390/ijerph110808137

Cairncross S, Feachem RG (1993) Environmental health engineering in the tropics: an introductory text, 2nd edn. Wiley, Chichester

Clasen T, Roberts I, Rabie T, Schmidt W, Cairncross S (2006) Interventions to improve water quality for preventing diarrhoea. Cochrane Database Syst Rev 3:CD004794. doi:10.1002/14651858. CD004794.pub2

Cronk R, Slaymaker T, Bartram J (2015) Monitoring drinking water, sanitation, and hygiene in non-household settings: priorities for policy and practice. Int J Hyg Environ Health 218(8):694–703. doi:10.1016/j.ijheh.2015.03.003

Cumming O, Elliott M, Overbo A, Bartram J (2014) Does global progress on sanitation really lag behind water? An analysis of global progress on community- and household-level access to safe water and sanitation. PLoS One 9(12):e114699. doi:10.1371/journal.pone.0114699

Dangour AD, Watson L, Cumming O, Boisson S, Che Y, Velleman Y, Cavill S, Allen E, Uauy R (2013) Interventions to improve water quality and supply, sanitation and hygiene practices, and their effects on the nutritional status of children. Cochrane Database Syst Rev 8:CD009382. doi:10.1002/14651858.CD009382.pub2

Ejemot RI, Ehiri JE, Meremikwu MM, Critchley JA (2008) Hand washing for preventing diarrhoea. Cochrane Database Syst Rev 1:CD004265. doi:10.1002/14651858.CD004265.pub2

Esrey SA, Potash JB, Roberts L, Shiff C (1991) Effects of improved water supply and sanitation on ascariasis, diarrhoea, dracunculiasis, hookworm infection, schistosomiasis, and trachoma. Bull World Health Organ 69(5):609–621

Fewtrell L, Kaufmann RB, Kay D, Enanoria W, Haller L, Colford JM Jr (2005) Water, sanitation, and hygiene interventions to reduce diarrhoea in less developed countries: a systematic review and meta-analysis. Lancet Infect Dis 5(1):42–52. doi:10.1016/S1473-3099(04)01253-8

Foster T (2013) Predictors of sustainability for community-managed handpumps in sub-Saharan Africa: evidence from Liberia, Sierra Leone, and Uganda. Environ Sci Technol 47(21):12037–12046. doi:10.1021/es402086n

Kayser GL, Moriarty P, Fonseca C, Bartram J (2013) Domestic water service delivery indicators and frameworks for monitoring, evaluation, policy and planning: a review. Int J Environ Res Public Health 10(10):4812–4835. doi:10.3390/ijerph10104812

Onda K, Lobuglio J, Bartram J (2013) Global access to safe water: accounting for water quality and the resulting impact on MDG progress. World Health Popul 14(3):32–44

Pickering AJ, Davis J (2012) Freshwater availability and water fetching distance affect child health in sub-Saharan Africa. Environ Sci Technol 46(4):2391–2397. doi:10.1021/es203177v

Rabie T, Curtis V (2006) Handwashing and risk of respiratory infections: a quantitative systematic review. Trop Med Int Health 11(3):258–267. doi:10.1111/j.1365-3156.2006.01568.x

Sobel J, Mahon B, Mendoza CE, Passaro D, Cano F, Baier K, Racioppi F, Hutwagner L, Mintz E (1998) Reduction of fecal contamination of street-vended beverages in Guatemala by a simple system for water purification and storage, handwashing, and beverage storage. Am J Trop Med Hyg 59(3):380–387

Stelmach RD, Clasen T (2015) Household water quantity and health: a systematic review. Int J Environ Res Public Health 12(6):5954–5974. doi:10.3390/ijerph120605954

The Lancet (2012) Progress in sanitation needed for neglected tropical diseases. Lancet 379(9820):978. doi:10.1016/S0140-6736(12)60412-7

UN (1977) Report on the United Nations water conference. United Nations, Mar de Plata

UN (2000) United Nations Millennium Declaration, resolution adopted by the General Assembly. United Nations, New York

UN (2003) Indicators for monitoring the millennium development goals: definitions, rationale, concepts and sources. United Nations, New York

UN (2005) UN millennium project task force on water and sanitation–health, dignity and development: what will it take? United Nations Millennium Project, London

UN (2011) UN reports improved access to safe drinking water, but poorest till lagging. United Nations, New York

UNDP (1990) New Delhi statement, adopted at the global consultation on safe water and sanitation for the 1990s. United Nations Development Programme, New Delhi

Waddington H, Snilstveit B, White H, Fewtrell L (2009) Water, sanitation and hygiene interventions to combat childhood diarrhoea in developing countries. International Initiative for Impact Evaluation, New Delhi

Wang X, Hunter PR (2010) A systematic review and meta-analysis of the association between self-reported diarrheal disease and distance from home to water source. Am J Trop Med Hyg 83(3):582–584. doi:10.4269/ajtmh.2010.10-0215

White GF, Bradley DJ, White AU (2002) Drawers of water: domestic water use in East Africa. 1972. Bull World Health Organ 80(1):63–73; discussion 61–62

WHO (1997) Guidelines for drinking-water quality: surveillance and control of community supplies, vol 3. World Health Organization, Geneva

WHO/UNICEF (2010) Rapid assessment of drinking water quality. World Health Organization United Nations Children's Fund, Sweden

WHO/UNICEF (2012a) Millennium development goal drinking water target World Health Organization. United Nations Children's Fund, Geneva

WHO/UNICEF (2012b) Progress on drinking water and sanitation: 2012 update. World Health Organization, Geneva/New York

Wright J, Gundry S, Conroy R (2004) Household drinking water in developing countries: a systematic review of microbiological contamination between source and point-of-use. Trop Med Int Health 9(1):106–117

WSSCC (2000) Vision 21: a shared vision for hygiene, sanitation and water supply and a framework fro action. Water Supply and Sanitation Collaborative Council, Geneva

# Chapter 6
# Mobilizing Payments for Water Service Sustainability

Johanna Koehler, Patrick Thomson, and Robert Hope

**Abstract** The current model of rural water service delivery is broken. Money flows down from donors and governments to install infrastructure but little reliable information on performance flows back. Increased use of handpump mapping exercises by survey teams may usefully identify handpumps working one day of the year but this leaves the remaining 99.7 % of any year unknown. The continuous monitoring of services is increasingly important if we are to know the real level of water services being enjoyed by rural communities, with the growing consensus on the Human Right to Water adding further impetus. For governments and donors, knowing whether investments deliver verifiable impacts over time rather than simply knowing that budgets have been spent, is transforming established thinking. Mobile networks provide an inclusive architecture to reduce the information asymmetry between investments and outcomes. Information alone is insufficient to make progress but it is necessary to track and improve accountable service delivery. Information can improve institutional performance and help define appropriate roles and responsibilities between communities, governments and donors to close the loop between well-meaning investments and quantifiable outcomes. Donors can demonstrate value-for-money, government and water service regulators can align performance with measureable outcomes, and communities can contribute to financial sustainability through user payments that are contingent upon service delivery. Using unique observational data from monitoring handpump usage in rural Kenya, we evaluate how dramatic improvements in maintenance services influence payment preferences across institutional, operational and geographic factors. Public goods theory is applied to examine new institutional forms of handpump management. Results reveal steps to enhance rural water supply sustainability by pooling maintenance and financial risks at scale supported by advances in monitoring and payment technologies.

**Keywords** Water pumps • Hand pumps • Payments • Service delivery

Portions of this chapter were first published in World Development (http://www.sciencedirect.com/science/article/pii/S0305750X15001291). Reprint allowed by Creative Commons Attribution License. Adaptations and updates provided by the authors.

J. Koehler (✉) • P. Thomson • R. Hope
Smith School of Enterprise and the Environment, University of Oxford,
South Parks Road, Oxford OX1 3QY, UK
e-mail: Johanna.Koehler@ouce.ox.ac.uk

© Springer International Publishing Switzerland 2016
E.A. Thomas (ed.), *Broken Pumps and Promises*,
DOI 10.1007/978-3-319-28643-3_6

57

## 6.1   Introduction

In this chapter, the Oxford University Smart Handpumps study is reviewed, in which cellular based sensor technology enabled improved pump servicing and unlocked community willingness to pay, thereby creating a closed loop feedback cycle aligning payments with service level performance. The chapter summarizes the arguments presented by Koehler et al. (2015) and Oxford/RFL (2014). In an independent peer review by Welle et al. (2015) assessing factors that affect success in rendering water services sustainable based on ICT-reporting across eight projects globally, the Smart Handpumps project is identified as the only rural water ICT initiative that fulfilled their three criteria for "success": (a) ICT reporting takes place, (b) ICT reports are processed by government or service provider, and (c) service improvement as a result can be observed (Welle et al. 2015).

## 6.2   Context

### 6.2.1   The Rural Water Challenge

Since the latter years of the Decade of International Drinking Water Supply and Sanitation, 1981 to 1990, community management of rural water supply has been advocated by international organizations, governmental and non-governmental alike (Briscoe and de Ferranti 1988; Carter et al. 1999; Churchill et al. 1987; Harvey and Reed 2004; Jimènez and Pèrez-Foguet 2010; Therkildsen 1988; Whittington et al. 2008). The empowerment of communities is based on the principles of participation, decision-making, control, ownership and cost-sharing (Briscoe and de Ferranti 1988; Lockwood 2004). However, despite the positive characteristics of community management, operations and maintenance have barely improved (Blaikie 2006; Lockwood 2004). Failure is largely blamed on poor planning and service delivery (Carter et al. 1999, 2010; The World Bank Water Demand Research Team 1993), limited community financing (Carter et al. 2010; Harvey 2007; Harvey and Reed 2004; Skinner 2009) and shortcomings in the institutional design of management models (Sara and Katz 2010; Whittington et al. 2008). Consequently, rural water supplies are in danger of falling into a spiral of decline in the post-construction phase (Rouse 2013). Adoption of simplified infrastructure asset management principles can increase cost-effectiveness and reduce interruptions in service (Boulenouar and Schweitzer 2015). Whilst maintaining community-based models, new approaches are therefore required which acknowledge the communities' inability to maintain their water supply without support in the long term (Harvey and Reed 2004; Lockwood 2004).

An enduring puzzle in achieving progress towards universal and reliable water service delivery in Africa is overcoming barriers to sustainable water user payments for community-managed handpumps (Harvey and Reed 2004). The non-functioning of one third of the handpumps in rural Africa (RWSN 2009) has resulted in an uncertain return on the USD 1.2–1.5 billion of infrastructure investments in the last two decades (Baumann 2009). Increasing water service coverage has failed to translate into a guarantee of reliable service delivery (Hope and Rouse 2013; Therkildsen 1988; Thompson et al. 2001). The long repair times that contribute to high handpump failure rates in rural Africa are essentially associated with weak payment systems (Foster 2013; Harvey 2007; RWSN 2009). Community management of water services has been widely identified as a dominant but failing model in rural water service delivery in Africa (Banerjee and Morella 2011; Hope 2014) with growing evidence that improved payment systems promote handpump sustainability (Foster 2013). Increasing opportunities to exploit the new, inclusive and low-cost mobile infrastructure offer new but untested approaches to accelerate and maintain reliable water services for the 273 million rural Africans without improved water coverage (Hope et al. 2012; WHO 2014). The policy implications are relevant to the post-2015 debate on the Sustainable Development Goals (SDGs) and may increase momentum for universal and sustainable water services within the framework of the Human Right to Water and Sanitation (UNGA 2010).

### 6.2.2   Demand and Service Level

Since the Dublin Principles of 1992 (ICWE 1992), the demand-responsive approach has provided the template for most rural water supply services. It focuses on both financial and managerial sustainability through participatory planning, informed choices, and cost recovery or cost-sharing arrangements (Sara and Katz 2010). It involves households in the choice of technological and institutional arrangements, while requiring them to pay for the service (Whittington et al. 2008). Communities rather than donors or governments make informed choices about the preferred service level, which is reflected in their willingness to pay. They also decide on service delivery mechanisms, operation and maintenance of services as well as the management of and accounting for funds and the degree to which the private sector is involved (Deverill et al. 2001; Lockwood 2004; The World Bank Water Demand Research Team 1993). To best serve the users' preferences, economic and social constraints are considered in the user group's institutional design.

   However, in practice the success of the demand-responsive approach can be thwarted through a lack of acceptability, feasibility, or the limited capacity of communities to sustain the chosen option (Skinner 2003). The failure of communities to speedily repair their handpumps results in longer-term non-functionality causing discontent amongst water users, who then look for alternatives and refrain from paying fees – a process that leads to a downward spiral in water services (Cross and Morel 2005). To counter such a downward development, supra-communal management options should be considered for rural water services recognizing the critical importance of the interface between a community-based model and the local community it is meant to serve (Blaikie 2006). Banerjee and Morella (2011) demonstrate that central, regional or local governments play a dominant role in all aspects of energy, road and water infrastructure provision across Africa. However, it is only in the area of providing and maintaining water services where local communities are given a leading role – precisely the area where Banerjee and Morella (2011) identify most challenges. Alternatives such as private rural water service providers are promoted by Kleemeier and Narkevic (2010), who argue for private firms or individuals to receive long-term government-let contracts to design, build or rehabilitate, operate and maintain water supplies within a defined geographical area (Kleemeier and Narkevic 2010). This demonstrates that innovative institutional models are required that are able to align demand and service levels.

### 6.2.3  Institutional Choices

Institutions, "the humanly devised constraints that structure political, economic and social interaction" (North 1991), evolve over time and are adapted to specific human needs. This study in part focuses on those institutions that have been created for the management of groundwater resources, and specifically for managing handpumps in rural areas. Due to its delineation of management systems along the lines of rivalry of consumption and exclusion, the theory of public goods, building on Samuelson (1964) is chosen for analyzing the institutional design at community level. Two versions of the theory are applied – Ostrom's (1990) understanding of common pool resources (CPRs) and Buchanan's (1965) definition of club goods. While the non-excludable and rivalrous CPR is a "natural or man-made resource system that is sufficiently large as to make it costly… to exclude potential beneficiaries from obtaining benefits from its use" (Ostrom 1990), the excludable and non-rivalrous club good determines a membership margin at "the size of the most desirable cost and consumption arrangement" (Buchanan 1965). Ostrom (1990) defines principles for robust common pool resource institutions, requiring clear institutional rules and solution mechanisms. Buchanan's (1965) criteria for the management of club goods expand on the public-private spectrum and emphasize consumption/ownership/membership arrangements. Consumption-sharing models, tariffs and membership levels are determined by the local communities according to their particular requirements to prevent "congestion".

If adapted to handpump management, the institutional design is a response to varying group preferences with implications for payment behavior: Some groups prefer higher payments at household level to be able to limit abstraction and usage levels by reducing the number of users, thereby organizing themselves as "handpump clubs" with a more exclusive membership; others prefer lower individual payments but with higher membership numbers to ensure that enough money is available to pay for maintenance bills, thus treating the pumps as common pool resources.

## 6.2.4  Geographic Challenges and Infrastructure Decisions

A problem specific to sub-Saharan Africa is that low population density encourages broad spatial distribution between handpumps and the clustering of systems around existing infrastructure (Harvey and Reed 2004). This implies high opportunity costs for users, often women, who have to walk long distances to the next-best pump alternative when their usual pump breaks (Van Houweling et al. 2012). As the most urgent demand tends to occur in areas of widely scattered pumps, geography appears to have an important impact on payment behavior. Another geographical aspect is the distance of handpumps to spare parts outlets, which impacts the reliability of service delivery (Harvey and Reed 2006). Similarly, in a study covering 25,000 pumps across three countries in sub-Saharan Africa, Foster (2013) found that distance from the district or county capital city is significantly associated with non-functionality of handpumps.

## 6.2.5  Mobile-Enabled Service Delivery

Mobile networks provide an inclusive architecture to reduce the information asymmetry between investments and outcomes. Though not progress in itself, information offers the opportunity to track and improve service delivery. It can also improve institutional performance and accountability by helping define effective roles and responsibilities between communities, governments and donors to close the loop between well-meaning investments and quantifiable outcomes. Donors can demonstrate value-for-money, government and water service regulators can produce measurable outcomes, and communities can contribute to financial sustainability through user payments that are contingent upon service delivery.

In our trial, GSM transmitters were securely fitted inside the handle of handpumps. The transmitter automatically sends data on handpump use via SMS over the mobile phone network. It is small and robust without moving parts with a specially designed antenna that fits discreetly to the handle. Installation is simple, enabling it to be retrofitted into existing pumps in the field or built into new pumps prior to deployment. The prototype smart handpump was tested in Lusaka in July 2011 with peer-reviewed results published early the following year (Thomson et al. 2012a). This led to the Kyuso study with the first installation in August 2012 preceding the full trial running from January through December 2013.

The transmitters have an algorithm that translates pump usage, measured by the movement of the handle, into an estimate of the volume of water produced by the pump to which it was fitted. Extensive tests in Zambia during the initial proof-of-concept phase (Thomson et al. 2012a) and further tests in Kenya generated a calibration that gave a volumetric output that was +/− 10 % of the observed output. However, there are many factors that will vary this accuracy across pumps and for this study pumps were not calibrated individually: the same calibration formula was used for all pumps. As such, liters pumped was primarily viewed as a proxy for pump usage, not as a direct statement of the volume of water used in a way that should be considered equivalent to a water meter.

## 6.3 Research Objectives, Study Site and Methodology

The primary research question is whether timely information of handpump failures can drive a maintenance model that leads to faster repairs, and, in turn, increase community willingness to pay for pump maintenance services.

The study site comprises Kyuso District, Kenya, (38° 10′E, 0° 35′S; 660–880 m elevation; 2446 km$^2$) located 267 km east of Nairobi. Based on Government of Kenya census data (2009), the population of 26,848 households is almost entirely rural (99 %) with 62 % living in absolute poverty – one of the highest rates in Kenya (KNBS 2006; Kenya 2009). Frequent droughts exacerbate the area's poverty by adversely affecting the farmers' major source of income from crop yields and livestock. The mean annual temperature ranges between 26 °C and 34 °C. The bi-modal rainfall pattern, with long rains from March to May and short, heavier rains between October and December, drives handpump usage patterns with pumps more heavily used in the dry season. An estimated 70 % of households rely on unimproved sources, such as ponds and rivers (Kenya 2009), which has negative health implications. Of the remainder, 30 % use wells or boreholes, which include 66 Afridev handpumps installed over the last 20 years that were included in the study.

These 66 pumps have been equipped with mobile-enabled transmitters reporting hourly pump usage to a central server via SMS (Thomson et al. 2012a). About half the pumps were "actively managed" and sent data automatically to the server. The others had usage data recorded for later analysis. The latter group were classed as "crowd-sourced", although in both cases, users were provided with contact information to call in case of breakdown. When a handpump failure was noted, a mechanic was dispatched immediately to assess and fix the problem. This service was provided free of charge on the assumption that a good service had to be demonstrated in order to establish the maintenance model was viable and build trust that a faster repair service was feasible, *prior* to any payment mechanism being introduced. We also examined the willingness to pay preferences of rural water users after experiencing the service for a one-year trial period.

During 2013, the year of the study, handpumps broke two times per year on average; however, the range was between zero and 11, which led to a high variation in repair cost ranging from USD 54 to USD 649 per pump per year with an average repair cost of USD 62. This unpredictability of pump failures and the variation in cost indicate that pooling payments across the District may afford the users higher security against water supply failure risk. Therefore, a supra-communal management structure was proposed to explore a mobile payment platform that could build on high (73 %) use of mobile money services in Kyuso District, the majority of this being Safaricom's M-Pesa service. In such a scheme all members would contribute monthly cash payments to be deposited into a designated mobile payments account by the water user committee treasurer. SMS messages would subsequently inform users that their fees have been received and deposited into the account, thus creating greater transparency and accountability for the user group. If pooled, even costly repairs can be covered following an insurance-based approach.

During handpump downtimes in the dry season, 77 % of households report using a non-pump alternative drinking water source, whereas 64 % use such sources during the wet season, which may cause seasonal shifts in pump revenue. Two major alternative sources in the area are Kiambere water pipeline and Ngomeni rock catchment, which provide piped water through kiosks (USD 0.02 per 20 l) for people living in the limited service area.

Within this study, four factors were hypothesized to be major influences on demand for a certain service level of rural water supply: handpump service reliability, handpump density, water use and water quality. The first three factors form the basis of the sampling framework and the analysis in this chapter. Water quality is not examined here but is a goal of further research in the site. The institutional framework depicts an organization of users whose preferences determine payment level and mode in order to achieve a certain service level supply (Fig. 6.1). The institutional design of the water user group is a key factor in achieving regular rural water user payments as it constitutes a link between the individual user and the supra-communal management structure in terms of personal involvement in the user group and willingness to pay.

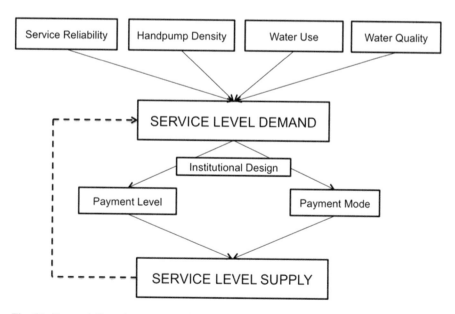

**Fig. 6.1** Factors influencing payments for rural water services (Koehler et al. 2015)

## 6.4   Baseline Survey

In order to test the model against the *status quo* and understand the choices and preferences of water users, a baseline survey was conducted in July 2012. This involved interviewing 124 voluntary respondents who were collecting water at 21 handpumps. Sampled respondents were mainly female (64 %) with an average age of 41 years. The average household size was 5.3 household members. Median adult equivalent expenditure was USD 313 per year, which represents two thirds (68 %) of the global poverty line of USD 1.25 per person per day. Handpumps provide the majority of households with their main drinking water source (59 %) and cooking, bathing and washing water (67 %) throughout the year. However, the dominant use of handpump water for households is for livestock watering (74 %). Almost nine in ten households (86 %) consider the water safe to drink though one in three claimed to treat the water by either boiling or chlorination.

### 6.4.1   Pump Breakdown Data

In the previous 12 months to the survey, 18 of the 21 handpumps (86 %) had experienced a failure. The average failure rate was just over two failures per year (range 0–10) with a total of 48 failures. The median repair time to fix a pump was six days with an average of 27 days (7 % of cases had a downtime of over a year). To generate these data multiple informants were asked and the average of their responses taken as the downtime. There was wide variation in responses but the aggregated data were consistent with available estimates. For example, a mapping survey of 440 Afridev handpumps in Kwale County by the project team in September 2013, estimated the average downtime of handpumps per breakdown at 37 days for functional pumps and 85 days for non-functional pumps. Average downtime estimates at 'functioning' handpumps will under-estimate the aggregate downtime figure as illustrated by the longer downtime in non-functioning handpumps in Kwale. Given the likely recall bias in estimating downtime by respondents we chose to use the figure of 27 days downtime per handpump per year as our baseline figure as it will likely be a conservative estimate.

Over two in five households (44 %) indicated they did not pay for water from their handpump. Among the majority that did pay a portfolio of overlapping payment approaches existed including a one-off membership fee, a monthly user fee and pay-as-you-go fees for drinking water containers or head of livestock. The most common payment modes were monthly fees, which were generally USD 0.56 per month (30 % of all paying households) or USD 1.1 per m³ water (Ksh 2 per 20 l container; 28 % of all paying households). The opaque approach to fee collection is reflected in the financial challenges when a handpump breaks and needs repairing. It was reported that there are sometimes sufficient funds (24 %), but more often there are not (36 %), or funds which only cover minor repairs (18 %).

Handpump failures are aggravated by delays in raising money in 40 % of cases with an average of 18 days to raise sufficient funds (median=7 days; range 1–180 days). Unprompted concerns about handpump management showed maintenance to be a key priority across a range of overlapping factors:

1. Repairs are too expensive (19 %)
2. Repairs take too long (17 %)
3. Handpump breaks too often (17 %)
4. Too many users (10 %)
5. Pump too far (8 %)
6. Water unsafe to drink (6 %)
7. Water fee too high (1 %)

### 6.4.2  User Preferences for Maintenance Service

As part of the survey a choice experiment tested handpump user preferences for alternative maintenance models. The experiment was orthogonally designed with ten pictorial cards that required choices across competing attributes of maintenance provider, maintenance level, payment mode, and payment level. A sample of 3540 observations was produced from usable data from 118 handpump users. Results identify community management of maintenance services as the least preferred option.

## 6.5  Experimental Design

### 6.5.1  Smart Handpumps Trial

Randomized Control Trials (RCT) are often viewed as an ideal experimental design as they eliminate potential selection biases and confounding factors will be randomized across groups. However, as is often the case in field trials, an RCT was not feasible for methodological, statistical and ethical reasons:

1. Random assignment of fee-paying and control handpumps would have likely co-located some controls with treatment handpumps, as a significant number of pumps were in close geographical clusters. This would have biased the results as control handpumps could have been abandoned in favour of treatment pumps after the first failure.
2. The sample size in the study area gave sufficient power, with two treatments of equal size, to show the level of effect that we predicted. However, if this had been two treatments and a control of equal sizes, the sample size would not have given sufficient power for statistical inference.

3. On ethical grounds, the local government (District Water Office) did not support treating rural water users differently, and thus having a control group was not an option.

Instead, the design had two treatments and included all the surveyed handpumps in the District. The split between the two treatments was not random, but was determined by mobile signal strength sufficient to reliably transmit SMS messages. This may have created a bias in terms of the ability of communities in the crowd-sourced treatment to call in case of breakdowns. Usually there was signal in the settlement which the pumps served, but commonly none at the handpumps as they were often located in a depression. Distance between treatments and the District Water Office was not significantly different (average 22 km). Observed water usage levels were not known before the trial though average usage level of the two treatments was found to be broadly similar (average 66 vs. 75 l per hour).

Handpumps that were non-functional for environmental reasons (e.g. a dry well that required further excavation) were not included in the analysis of maintenance response times. There were four repairs after which the same pump needed to be immediately repaired again. In these cases the repairs were collapsed to a single repair, with the overall time to repair used in the calculations.

## 6.5.2 Willingness to Pay Methodology

To explore community preferences to institutional barriers and opportunities, focus group discussions (FGD) were administered by two native Kikamba speakers supported by the lead author. A total of 63 field days were spent in 66 handpump communities in the periods June/July and November/December 2013. In total 639 participants were included in this process.

In the first phase 32 of the 66 handpumps were systematically sampled according to the following criteria: density category, experience of service and level of usage. In round two the remaining 34 handpumps were sampled. Follow-up focus groups were conducted at the community pumps. In the first round groups were divided by gender. This methodology was chosen because women might be reluctant to state their own preferences in the presence of men (The World Bank Water Demand Research Team 1993). The follow-up was conducted in mixed groups because groups had met in the interim to discuss the proposals. Participants ranged in age from 20 to 80 years and represented both users with and without mobile phones. FGD methods included mapping the water user community with alternative sources, a seasonal calendar, and a timeline on handpump maintenance (Narayanasamy 2009).

A group willingness to pay activity was designed to identify how much each water user group would be willing to pay for a continuation of the experienced maintenance service at the conclusion of the free maintenance trial in December 2013. No maintenance service standard was guaranteed as the aim was to understand individual community preferences based on their experience of the service. Community experiences varied from no maintenance response, as 30 % of hand-pumps did not fail in 2013, through to communities having had their handpump repaired on at least one occasion. A willingness to pay design was chosen as a means to initiate community debate on payments collectively. Without rehearsing the extensive literature critiquing willingness to pay studies conducted with individuals (Hensher et al. 2005; Merrett 2002) or collectively (Wiser 2007), well-established biases (strategic, protest vote, anchor) and limitations (temporal invariance, intra-household dynamics, social dynamics, computation) are acknowledged. Davis' study (2004) on the effects of the mode of data elicitation on results obtained in demand-assessment research demonstrates that the explanatory power is highest in a combination of focus groups and subsequent self-administered questionnaires, which she largely attributes to additional time for contemplation. As in our case group decision-making was the objective for a standard payment level per user group, we replaced the questionnaires by follow-up focus groups leaving time for each group to reach consensus.

Importantly, the research team did not suggest minimum payment nor did it prescribe a payment system (from equality to a sole benefactor) but supported the group discussion with a view to engage quieter members actively but respectfully in an inclusive discussion. The water user committee members attending focus group discussions were also interviewed separately regarding current water user committee management, thus informing the discussion on excludability. Additional interviews with user group members provided insight into relevant group dynamics.

Data were analyzed in three steps: firstly, the quantitative willingness to pay was analyzed according to the themes developed in the sampling framework. The statistical program SPSS, version 22, was used for statistical tests. Secondly, focus group transcripts were coded according to themes, which added narrative to the quantitative findings (Miles and Huberman 1994). Thirdly, the analysis of excludability through a ranking system determined management types as common pool resources, club goods or privately managed pumps.

## 6.6 Evidence for Institutional Reform

### 6.6.1 Service Level

A key outcome was whether timely information of handpump failures can drive a maintenance model that leads to faster repairs. To this end we recorded when a maintenance alert was first raised and when the repair was actually completed, using the difference to indicate the time the handpump was not working and hence the effectiveness of the new maintenance model. The following table shows the range of downtimes for the two treatments (active, crowd) in the trial and the baseline (Table 6.1).

**Table 6.1** Handpump repair interval (Oxford/RFL 2014)

| Group | | Mean days to repair | Median days to repair | Increase in chance of repair within two days vs. baseline[a] | Repairs |
|---|---|---|---|---|---|
| Baseline survey | | 27 | 6 | – | 48 |
| Study | All repairs | 2.6 | 1 | 4.3 | 111[b] |
| | Actively managed | 2.0 | 1 | 4.8 | 74 |
| | Crowd sourced | 3.7 | 3 | 3.2 | 37 |

[a]Risk Ratio or Relative Risk calculated with respect to the baseline case. All p-values are <0.001
[b]This figure excludes instances when the repair time was not related to the maintenance service (e.g. the local community had to deepen the well before additional pipes and rods could be added by the mechanic)

The results provide evidence of significant improvements under the new maintenance model:

- Pump outage times drop by *an order of magnitude* from a mean of 27 days to under three.
- Eighty-nine percent of repairs were completed within five days, rising to 95 % for the actively managed group.
- A handpump is over *four times more likely* to have been repaired *within two days* than before the trial.
- The actively-managed handpumps were 50 % more likely to have been fixed within two days than the crowd-sourced handpumps.

## 6.6.2   Willingness to Pay

Poor service levels appear to be the most important barrier to sustaining water user payments. Increased reliability, enabled through mobile monitoring, constitutes a critical component of demand as it affects other preferences. For this purpose, monthly payment levels per household for the time before the service started were compared with willingness to pay levels after the users had experienced the service ($n = 46$) including those private pumps that would join the payment model. The increase is fivefold from USD 0.2 to USD 1 per household per month. This can be related to the fact that the new level of service produced a tenfold decrease in handpump downtime from 27 to 2.6 days on average over the one-year study period, which represents an order of magnitude improvement found to be critical in the baseline survey (Hope 2014). Moreover, the number of handpump groups intending to contribute monthly – rather than making post-breakdown payments – increased threefold.

The findings suggest that payments are contingent on service delivery. The aspects of service delivery that were most valued by the water users were the speed of service (77 %), the quality of the service (54 %) and the knowledge that the service is guaranteed (31 %). The focus group participants also endorsed mobile payments as an acceptable payment mode, especially as mobile payments are already used for remittances by at least one member in each focus group. The average monthly willingness to pay for a mobile-enabled service at all 66 pumps is USD 0.92 per household; the average group willingness to pay is USD 21 per month across all pumps. Of the sample of 66 handpumps, 70 % required at least one repair in 2013, with 63 % of broken handpumps requiring more than one repair. The average cost of each repair was USD 62. If the stated willingness to pay of all pump user groups reflected the actual future payment collected, this would raise sufficient revenue to have covered all repair costs in 2013; however, if communities chose not to pool revenue, 43 % of communities would not have met their individual costs. Equal monthly payments and equal cost-sharing are deemed universally important. With the given level of acceptance and use of mobile phones in Kyuso, a mobile-enabled service delivery model is socially acceptable and familiar as well as practical and efficient.

Higher revenues – expressed here as greater willingness to pay – can lead to improved pump maintenance triggering greater benefits for users. The spiral of decline and discontent among users leading to non-payment and long-term pump non-functionality can be reversed through an effective maintenance system that facilitates demand for higher service levels (The World Bank Water Demand Research Team 1993). Translating this willingness to pay into actual payments requires strong institutions with enforcement mechanisms. Madrigal et al. (2011) point to the significance of a set of working rules enforced by the local communities. At the same time, acknowledging that different institutional arrangements may be chosen by different user groups - be that common pool resource groups (Ostrom 1990) or clubs (Buchanan 1965), depending on the chosen level of excludability - and allowing those preferences to be satisfied, reinforces institutional stability (Olson 1965; Ostrom 2010). In this study, the average membership size of exclusive clubs (27 members) is 43 % smaller than that of more inclusive CPR groups (47 members). Excludability is meant to prevent queuing, wear on the pump, overabstraction and potential rationing of the resource, while greater inclusivity allows lower membership fees. The water user committee plays an important role in administering rules and regulations that define the exclusivity of the group. A tighter organizational structure seems to be related to higher demand for water, as club handpumps have 57 % higher usage levels than CPR groups. Moreover, the application of public goods theory to the institutional design of user groups reveals that the more exclusive handpump clubs show a 43 % higher average willingness to pay per member per month (USD 1.03) than more inclusive groups classified as common pool resource groups (USD 0.72).

Geographic factors also influence rural water user payments, as the existence of alternative sources is likely to reduce the willingness of users to pay for the operation and maintenance of a certain pump from which they can easily switch to another one. Thus, handpump density has implications for operational management and investment planning as well as for the institutional design of the user groups. While the average group size for single pumps is 43 household members, it is 27 for clusters. The willingness to pay level is 47 % higher at single pumps at the household level and 2.6 times higher at the level of user groups. Given similar population density, handpump clustering is thus at best an inefficient distribution of resources and at worst a counter-productive planning decision (Narayan 1993).

Even beyond willingness to pay, a study by Ali et al. (2014) finds that satisfaction with public service provision promotes a tax-compliant attitude among Kenyans and other sub-Saharan Africans (Ali et al. 2014). This may indicate that willingness to pay will eventually translate into actual payments as the new, improved service level is recognized. Altogether, the measures discussed above have profound implications for the operational and institutional challenges of community management of handpumps.

## 6.7 Operationalizing and Institutionalizing Rural Water Services and User Payments

### 6.7.1 Harnessing Mobile Technology for Monitoring and Payment

Mobile monitoring and mobile payments have the potential to improve traditional payment systems with benefits for both service provider and water user. For the former, it provides an effective monitoring system that "would be alert to all credible problems and notify maintenance responses in a timely and constant manner" (Thomson et al. 2012b), thus not only enabling fast repairs but also contractual oversight. For the latter, benefits include a more transparent financial system and a higher level of water security through regular repairs (Hutchings et al. 2012). While mobile monitoring facilitates a hitherto impossible alignment of service delivery with user level demand through monitoring functionality and abstraction, mobile payments facilitate direct financial flows back to the maintenance service provider (Fig. 6.1). Mobile technology could therefore act as a conduit for reliable information and financial flows, thus achieving the central objective of strengthening handpump sustainability while increasing financial transparency and security. The service provider may achieve a better understanding of the financial capacity of water user groups while users can monitor their management committees through feedback loops, which would counter potential mismanagement of handpump finances. Without strict group level enforcement measures, the entire group may lose interest in fee collection (Harvey and Reed 2004).

Mobile technology is not a panacea for the rural water supply problems of Kenya or other countries. There are numerous obstacles impeding the successful delivery of a mobile-enabled service, including the lack of signal and electricity for recharging mobile phones, together with operational problems of crowd-sourcing (Daraja 2012). However, these technical challenges are surmountable with infrastructure coverage and mobile subscription levels continuously increasing. Technology is an *enabler* that creates the opportunity for novel management models, which were not previously possible; yet it will not train and equip mechanics, enforce agreed payment levels, or conduct a spare parts inventory check. Overall, success is contingent upon the willingness of the people to participate. "Getting the human side of things right… [is] much harder than making the technology work" (Daraja 2012). Nevertheless, this study has shown that, by aligning rural water supply systems to the service level demand formed through socio-economic preferences and translated into the institutional design of user groups, the financial sustainability of community handpumps may be improved. Mobile technology is a useful tool for aligning supply and demand through better information availability – but only if the institutional structure at the community and supra-communal level are sufficiently robust to nurture and exploit its full potential.

## 6.7.2   Developing an Output-Based Payment Framework

Overcoming the barriers to rural water user payments is an essential step in the global drive towards achieving the water targets of the sustainable development agenda. An output-based payment model represents a new framework for donor and government behavior in Kenya and other African countries (Fig. 6.2) within the wider initiatives on results-based payment approaches (DFID 2014; Hope 2014). This cycle of improved service delivery constitutes the first building block at the sub-national level where finances flow from communities to a

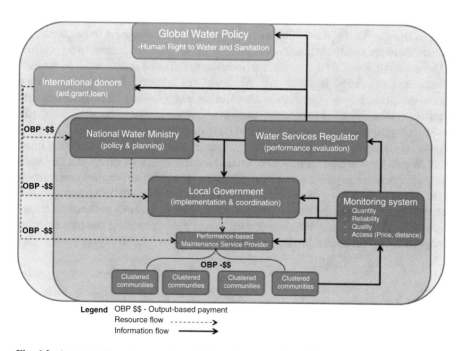

**Fig. 6.2**  An output-based payment model of rural water services (Koehler et al. 2015)

performance-incentivized maintenance service provider, who in turn is monitored and regulated by local and national governments. The national rural water regulatory system documents existing and new investments by environmental, technical and operational indicators, thus providing a valuable resource for monitoring and regulating investment behavior and outcomes at scale. Regular information on performance and user payments can drive a results-based financing mechanism that supports the provision of basic public services. This can be facilitated by delegating the delivery of outputs, such as a functioning maintenance service, to a third party in exchange for the payment of a subsidy upon delivery of specific outputs. It can thus address a potential funding gap between the cost of service delivery and the beneficiaries'ability and willingness to pay the full amount of user fees for the service (World Bank 2014; IDA 2009). Hence, this system can continuously inform national government goals and priorities while supporting global water policy approaches.

## 6.8   Study Limitations

A number of limitations are identified in this study. First, the study site is in one district in rural Kenya with a particular hydro-climatic, geological, social and political landscape; no claim is made to generalize the findings though there is confidence in internal validity. Second, the free maintenance service may have biased informant responses on payment levels upwards, as the respondents had been beneficiaries of this service for a significant period prior to the willingness to pay study. As noted, there are significant methodological concerns with willingness to pay studies (Davis 2004). Aware of the issues, we have attempted to be conservative in the estimates and associated implications. We could not address all the broader scope socio-cultural factors affecting willingness to pay of users. Third, insufficient resources were available to conduct analysis of environmental variation (hydrogeology, recharge, water quality) or technical components (installation quality, depth of well), which may confound some of the results. Future work aims to include natural and human-related contamination to understand the extent to which this key variable affects water payment behavior. Fourth, the research team worked closely with but independently from the District Water Office and its staff. While government support was instrumental in the research, we acknowledge such collaboration may have affected community behavior despite enforcing strict ethical and human informant measures on confidentiality and anonymity.

## 6.9   Conclusion

Understanding operational, geographic and institutional barriers of rural water user payments contributes to developing an innovative, output-based payment model for rural water services in Africa. The real test is if users support the introduction of a new payment system in the long term, which acknowledges the higher value for money that the new maintenance service system creates. This research indicates that such reforms are supported by the communities if reliable services are delivered. The findings offer pathways towards the water target of the post-2015 sustainable development agenda promoting, *inter alia,* universal and sustainable access to safe drinking water and raising service standards, as well as robust and effective water governance with more effective institutions and administrative systems (UN 2014). It demonstrates the need for continuous monitoring of rural water services, as well as suggesting strategies for achieving this. Water service performance data are key to defining a baseline and measuring progress towards sustainable services at the local level, to operationalizing a maintenance service provider model at the supra-communal level and testing an output-based payment model at the national and international levels. The Government of Kenya's Water Services Regulatory Board (WASREB) acknowledges the importance of such performance data "enabling WASREB to ensure that satisfactory performance levels are achieved and maintained, and enhancing transparency and accountability within the rural sector" (WASREB 2014). Thus, the data can support and monitor national policy goals that promote progress towards universal access and more reliable improved water services for the rural poor.

## 6.10 Post-Script

In 2015 a private maintenance service provider, named FundiFix Ltd., started operating a maintenance service, under a joint UNICEF and Oxford University research project. The enterprise developed a flexible tariff structure based on usage with preferential rates for institutions such as schools. The prepaid mobile payment system provides feedback messages to community members for improved accountability and transparency. While it is too early to present final results, an 89 % revenue collection was achieved in the first 6 months (Oxford/RFL 2015). However, it is also evident that environmental factors such as high salinity or low water levels in certain wells, are an important barrier for communities signing up for the maintenance contract. Such issues will not be resolved by a rapid repair service, greater oversight or financial innovation. This underlines the fact that water services and water resources are inextricably linked, and that the former cannot be universally delivered without the latter being understood and attended to.

## References

Ali M, Fieldstad OH, Siursen IH (2014) To pay or not to pay? Citizens' attitudes toward taxation in Kenya, Tanzania, Uganda, and South Africa. World Dev 64:828–842
Banerjee S, Morella E (2011) Africa's water and sanitation infrastructure. The World Bank, Washington, DC

Baumann E (2009) May-day! May-day! Our handpumps are not working!, vol perspectives no 1. Rural Water Supply Network, Switzerland

Blaikie P (2006) Is small really beautiful? Community-based natural resource management in Malawi and Botswana. World Dev 34(11):1942–1957

Boulenouar J, Schweitzer R (2015) Infrastructure asset management for rural water supply. IRC, The Hague

Briscoe J, de Ferranti D (1988) Water for rural communities. Helping people help themselves. World Bank, Washington, DC

Buchanan JM (1965) An economic theory of clubs. Economica (New Ser) 32(125):1–14

Carter R, Tyrell SF, Howsam P (1999) Impact and sustainability of community water supply and sanitation programmes in developing countries. J Chart Inst Water Environ Manage 13:292–296

Carter R, Harvey E, Casey V (2010) User financing of rural hand-pump water services. Paper presented at the IRC symposium 2010: pumps, pipes, and promises. IRC, The Hague

Churchill A, de Ferranti D, Roche R, Tager C, Walters A, Yazer A (1987) Rural water supply and sanitation: time for a change. World Bank discussion paper 18. World Bank, Washington, DC

Cross P, Morel A (2005) Pro-poor strategies for urban water supply and sanitation services delivery in Africa. Water Sci Technol 51(8):51–57

Daraja (2012) The failure of Maji Matone, Phase 1. Daraja. http://blog.daraja.org/p/failure.html. Accessed 10 Aug 2013

Davis J (2004) Assessing community preferences for development projects: are willingness-to-pay studies robust to mode effects? World Dev 32(4):655–672

Deverill P, Bibby S, Wedgwood A, Smout I (2001) Designing water and sanitation projects to meet demand in rural and peri-urban areas – the engineer's role. Interim report. World Bank, Washington, DC

DFID (2014) Sharpening incentives to perform: DFID's strategy for payment by results. Department for International Development, London

Foster T (2013) Predictors of sustainability for community-managed handpumps in sub-Saharan Africa: evidence from Liberia, Sierra Leone, and Uganda. Environ Sci Technol 47(21):12037–12046. doi:10.1021/es402086n

Harvey P (2007) Cost determination and sustainable financing for rural water services in sub-Saharan Africa. Water Policy 9(4):373

Harvey P, Reed R (2004) Rural water supply in Africa: building blocks for handpump sustainability. Loughborough University: Water, Engineering and Development Centre, Loughborough

Harvey PA, Reed RA (2006) Community-managed water supplies in Africa: sustainable or dispensable? Community Dev J 42(3):365–378

Hensher D, Shore N, Train K (2005) Households' willingness to pay for water service attributes. Environ Resour Econ 32(4):509–531

Hope RA (2015) Is community water management the community's choice? Implications for water and development policy in Africa. Water Policy 17(4):664–678

Hope R, Rouse M (2013) Risks and responses to universal drinking water security. Philos Trans A Math Phys Eng Sci 371(2002):20120417. doi:10.1098/rsta.2012.0417

Hope R, Foster T, Thomson P (2012) Reducing risks to rural water security in Africa. Ambio 41(7):773–776. doi:10.1007/s13280-012-0337-7

Hutchings M, Dev A, Palaniappan M, Srinivasan V, Ramanathan N, Taylor J (2012) Mobile phone applications for the water, sanitation, and hygiene sector. Pacific Institute and Nexleaf Analytics, Oakland

ICWE (1992) The Dublin Statement and Report of the Conference. In: ICWE (ed) International conference on water and the environment: development issues for the 21st century, Dublin, 26–31 January 1992. ICWE, Dublin

IDA (2009) IDA15 mid-term review: a review of the use of output-based aid approaches. International Development Association, Global Partnership on Output-Based Aid, Washington, DC

Jimènez A, Pèrez-Foguet A (2010) Challenges for water governance in rural water supply: lessons learned from Tanzania. Water Resour Dev 26(2):235–248

Kenya (2009) Kenya national census 2009. Government of Kenya, Nairobi

Kleemeier E, Narkevic J (2010) Private operator models for community water supply. World Bank: Water and Sanitation Program, Nairobi

KNBS (2006) Kenya integrated household budget survey. Kenya National Bureau of Statistics, Nairobi

Koehler J, Thomson P, Hope R (2015) Pump-priming payments for sustainable water services in rural Africa. World Dev 74:397–411

Lockwood H (2004) Scaling up community management of rural water supply. Thematic overview paper, International Water and Sanitation Centre, Den Haag

Madrigal R, Alpízar F, Schlüter A (2011) Determinants of community-based drinking water organizations. World Dev 39(9):1663–1675

Merrett S (2002) Deconstructing households' willingness-to-pay for water in low-income countries. Water Policy 4(2):157–172

Miles MB, Huberman AM (1994) Qualitative data analysis: an expanded sourcebook. Sage, London

Narayan D (1993) Participatory evaluation: tools for managing change in water and sanitation. The World Bank, Washington, DC

Narayanasamy N (2009) Participatory rural appraisal: principles, methods and application. Sage, London

North DC (1991) Institutions. J Econ Prospect 5(1):97–112

Olson M (1965) The logic of collective action: public goods and the theory of groups. Harvard University Press, Cambridge

Ostrom E (1990) Governing the commons: the evolution of institutions for collective action. Cambridge University Press, Cambridge

Ostrom E (2010) Analyzing collective action. International Association of Agricultural Economists, Milwaukee

Oxford/RFL (2014) From Rights to Results in Rural Water Services - Evidence from Kyuso, Kenya. Water Programme, Working Paper 1, Smith School of Enterprise and the Environment, Oxford University, UK

Oxford/RFL (2015) Financial Sustainability for Rural Water Services – Evidence from Kyuso, Kenya. Water Programme, Working Paper 2, Smith School of Enterprise and the Environment, Oxford University, UK

Rouse M (2013) Institutional governance and regulation of water services. IWA Publishing, London

RWSN (2009) Myths of the rural water supply sector. RWSN perspective no. 4. Rural Water Supply Network, Gland

Samuelson PA (1964) Economics: an introductory analysis. McGraw-Hill, New York

Sara J, Katz T (2010) Making rural water supply sustainable: report on the impact of project rules. UNDP-World Bank, Washington, DC

Skinner B (2003) Small-scale water supply: a review of technologies. ITDG Publishing, London

Skinner J (2009) Where every drop counts: tackling rural Africa's water crisis. IIED briefing papers. IIED, London

Therkildsen O (1988) Watering white elephants? Lessons from donor funded planning and implementation of rural water supplies in Tanzania. Scandinavian Institute of African Studies, Uppsala

Thompson J, Porras I, Tumwin J, Muiwahuzi M, Katui-Katua M, Johnstone N, Wood L (2001) Drawers of water II: 30 years of change in domestic water use and environmental health in East Africa. International Institute for Environment and Development, London

Thomson P, Hope J, Foster T (2012a) GSM-enabled remote monitoring of rural handpumps: a proof-of-concept study. J Hydroinf 14(4):29–39

Thomson P, Hope R, Foster T (2012b) Is silence golden? Of mobiles, monitoring, and rural water supplies. Waterlines 31(4):280–292

The World Bank Water Demand Research Team (1993) The demand for water in rural areas: determinants and policy implications. World Bank Res Obs 8(1):47–70

UN (2014) A post-2015 global goal for water: synthesis of key findings and recommendations from UN-water. UN Water, New York

UNGA (2010) Resolution 64/292: the human right to safe drinking water and sanitation. United Nations General Assembly, New York

Van Houweling E, Hall RP, Diop AS, Davis J, Seiss M (2012) The role of productive water use in women's livelihoods: evidence from rural Senegal. Water Altern 5(3):658–677

WASREB (2014) IMPACT – a performance review of Kenya's water services sector 2012–2013. Government of Kenya, Water Services Regulatory Board, Nairobi

Welle K, Williams J, Pearce J, Befani B (2015) Testing the waters – a qualitative comparative analysis of the factors affecting success in rendering water services sustainable based on ICT-reporting. IDS and Wateraid, Brighton

Whittington D, Davis J, Prokopy L, Komives K, Thorsten R, Lukacs H, Wakeman W (2008) How well is the demand-driven, community management model for rural water supply systems doing? Evidence from Bolivia, Peru and Ghana. BWIP working paper, 22. BWIP, Manchester

WHO (2014) Progress on drinking water and sanitation – 2014 update. World Health Organization and UNICEF, Geneva

Wiser R (2007) Using contingent valuation to explore willingness to pay for renewable energy: a comparison of collective and voluntary payment vehicles. Ecol Econ 62(3–4):419–432

World Bank (2014) Applying results-based financing in water investments. Water papers. The World Bank, Washington, DC

# Chapter 7
# Enabling Ecological Restoration Through Quantification

**Alex H. Johnson and Joe S. Whitworth**

**Abstract** The Freshwater Trust, a conservation non-for-profit headquartered in Portland, Oregon, has been working since 1983 on solutions to address river impairment in ways that are replicable and scalable to meet the today's environmental challenges. The methods and tools that The Freshwater Trust uses to return degraded waters to healthy waters are founded on the principles of measuring, or quantifying, environmental outcomes and tracking effectiveness to maximize ecological uplift. Current restoration actions include planting streamside, or riparian, forests to provide shade to streams and stabilize banks; placing large wood structures instream to provide habitat for spawning and rearing habitat; restoring flow to dewatered streams; and reconnecting streams to closed floodplains.

**Keywords** Watershed • Restoration • Clean water act • Water credits

## 7.1 The Pace and Scale of Restoration

Billions of dollars are spent each year to restore degraded rivers throughout the United States. Despite these efforts, 55 % of U.S. rivers remain unhealthy. There are myriad examples of freshwater resources in the U.S. showing signs of distress:

- The state of California is facing one of the most severe droughts on record (California 2015).
- The Mississippi, the world's fourth longest river, has become choked by fertilizers that create an enormous dead zone covering as many as 5400 mile$^2$—roughly the size of Connecticut (EPA 2015).
- The Colorado River has not reached the ocean since the 1990s due to over-allocation and instead ends at dry land (HCN 2014).

A.H. Johnson (✉) • J.S. Whitworth
The Freshwater Trust, 65 SW Yamhill St # 200, Portland, OR 97204, USA
e-mail: alex@thefreshwatertrust.org

© Springer International Publishing Switzerland 2016
E.A. Thomas (ed.), *Broken Pumps and Promises*,
DOI 10.1007/978-3-319-28643-3_7

- The Chesapeake Bay, once called the most productive estuary in the world, has remained on the U.S. Environmental Protection Agency's "dirty water list" since the late 1990s (CBF 2014).
- Lake Erie's burgeoning harmful algal bloom is predicted to be one of the most severe in 2015 (NOAA 2015).

Given the compromised state of freshwater systems, the U.S. finds itself at a critical juncture. The time has come to restore these systems at a rate and scale that outpaces existing and anticipated pressures. These pressures include the rapid rise in global population, increased demand for food production, and extreme weather events—all of which cause disruptions to clean and reliable sources of freshwater, and result in increased water scarcity and security issues, exposure to environmental pollution, and the destruction of valuable ecosystems and the resiliency that they provide.

However, fixing rivers is achievable. One way to more effectively allocate resources to improve watershed health is by using newly developed metrics and technologies to select restoration sites based on their ability to provide the greatest ecological uplift. Uplift is defined as the difference between degraded baseline conditions and future restored conditions. Understanding the potential environmental benefits of projects in advance will allow practitioners to focus on areas where restoration actions will have the greatest impact, and will make it possible to fix more rivers, faster.

The Freshwater Trust, a conservation non-for-profit headquartered in Portland, Oregon, has been working since 1983 on solutions to address river impairment in ways that are replicable and scalable to meet the today's environmental challenges. Formed by a merger of two independent nonprofits, Oregon Trout and the Oregon Water Trust, The Freshwater Trust benefits from its background as the first water trust in the U.S. and an implementer of large-scale habitat projects. It is now actively working in Oregon, Idaho, California and Colorado to develop freshwater conservation and restoration programs. The methods and tools that The Freshwater Trust uses to return degraded waters to healthy waters are founded on the principles of measuring, or quantifying, environmental outcomes and tracking effectiveness to maximize ecological uplift. Current restoration actions include planting streamside, or riparian, forests to provide shade to streams and stabilize banks; placing large wood structures instream to provide habitat for spawning and rearing habitat; restoring flow to dewatered streams; and reconnecting streams to closed floodplains. The Freshwater Trust also works with regulated entities such as wastewater treatment plants and hydropower operators to meet regulatory compliance through river restoration, through a markets-based concept called water quality trading, discussed in detail in this chapter.

Healthier freshwater resources not only play a crucial role in the natural ecosystem, but they also allow for the continued growth and long-term resiliency of the world's population, commerce, and agriculture.

## 7.2  Making It Count: The Case for Quantifying Water

While the quantity of water on Earth has not changed, the timing, placement and condition of the water's flow has changed. If all of the world's water were poured into a gallon jug, only a tablespoonful of it would be freshwater (Shiklomanov 1999). That tablespoon of freshwater represents less than 3 % of the world's water resources (Shiklomanov 1999), and its quantity and quality have not been measured or tracked well or correctly.

### 7.2.1  Accounting of Natural Resources

The environment is the ultimate closed-loop system. However, until only recently, economic models have assumed an abundant supply of natural resources, including freshwater, without accounting for the full costs and impacts of economic choices.

For example, there's no precise way to calculate how much water goes into making the hamburgers sold by American fast-food chains. The retail price listed on the menu doesn't provide the whole picture. The full cost of food production—including raising cattle and growing corn to feed the cattle and the impacts of both on the environment—is not reflected in the price.

In the U.S., the inability to properly understand and account for the tradeoffs between consuming and restoring ecosystems has impaired water quality. These impacts are perhaps most clearly expressed in the 8000-square-mile dead zone created each year by agricultural runoff in the Gulf of Mexico. After World War II, the "green revolution" in agriculture dramatically increased production of grains and row crops through new technologies, but it did not adequately consider the capacity of freshwater systems to continuously absorb the accumulated pollution (Whitworth 2015a).

These negative environmental impacts were not generally intentional, but they also weren't considered. Many in the agriculture industry lacked the tools and economic incentive to properly account for environmental impacts. The passage of the Clean Water Act in the U.S. in 1972 helped to spur thousands of successful pollutant reduction and habitat improvement projects. Related developments included the work to define the benefits and services provided by ecosystems (Ehrlich and Ehrlich 1981) to define the "ecological economics" of ecosystem services (Costanza 1991); and to show how impacts to ecosystems affect the security, livelihood, and health of society (WRI 2005).

## 7.2.2 Quantifying Conservation

A more recent method of accounting for environmental impacts is quantified conservation, a data-driven approach for quantifying the financial and ecological effects of long-term restoration actions within a regulatory and economic framework (Whitworth 2015a). Building on the foundation laid by the earlier environmental movement, quantified conservation uses a sophisticated set of tools to bring about measurable improvements to ensure both a healthy environment and a thriving economy. Quantified conservation is a framework built on five business principles (Whitworth 2015b):

- **Situational awareness** to provide an objective understanding of the real-time environmental problems faced;
- **A focus on outcomes** that define the results sought;
- **Innovation and technology** to achieve desired outcomes at the pace and scale required for success;
- **Data and analytics** to prioritize the environmental projects that have the most impact, measure results, and monitor progress; and
- **Gain-focused investments** that maximize every dollar spent by tying public and private investments to measurable gains achieved for the environment.

Using emerging environmental science and technology, quantified conservation shifts the focus from process-based restoration ("how") to outcome-based restoration ("what"). A demonstration of the shift in conservation thinking toward the accountable quantified conservation model is summarized below (Table 7.1).

The shift in approach from "no net loss" to "net environmental gain" represents the next generation of conservation. To achieve this shift, all participants will be required to work in new ways. Regulators must adopt new policies that use the principles of quantified conservation in managing ecosystem impacts and restoration. Restoration project managers must shift from practice-based measures to outcome-based measures in their work. Funders must require that work be measured and tracked over time. Landowners must rethink on-farm practices in order to reasonably maintain productive lands for the long term, with minimal ecosystem impacts.

**Table 7.1** Traditional conservation methods compared to quantified conservation method

| Traditional conservation | Next-generation quantified conservation |
| --- | --- |
| Problem-focused | Solution-focused |
| Procedure-based | Outcome-based |
| Defensively focused | Action-oriented |
| Advocacy/litigation-oriented | Ecosystem services-oriented |
| "No net loss" | "Net environmental gain" |
| Generalized good effort | Scalable, replicable and measured results |

**Table 7.2** Basic example from the past: planting project *without* uplift metrics

| Acres | Trees planted | Total cost |
|-------|---------------|------------|
| 10 | 5000 | $50,000 |

**Table 7.3** Example of quantified conservation: planting project *with* uplift metrics

| Acres | Trees planted | Total cost | Kilocalories/day of solar load avoided | Pounds/year of phosphorus reduced | Weighted linear feet of salmon habitat restored |
|-------|---------------|------------|----------------------------------------|-----------------------------------|-------------------------------------------------|
| 10 | 5000 | $50,000 | 50,000,000 | 50 | 100 |

## 7.2.3 Applying Quantified Conservation

Not all parts of a river or stream are created equal. Using the quantified conservation method, restoration professionals can focus efforts on the actions and in the places where they will result in the greatest ecological uplift.

For example, if the objective is to reduce solar loading, then professionals can calculate the *benefit* of planting trees by quantifying the solar heat that will be blocked by those trees at maturity. Benefits can be expressed in terms of kilocalories per day of solar load avoided on a given stream. Calculations also exist to determine other benefits from healthy stands of trees, including reductions in pollutants (such as phosphorus carried from fields in runoff and absorbed by tree roots) and increases in habitat for fish species. This quantified conservation approach provides measurable metrics of a project's outcome (Tables 7.2 and 7.3).

The inclusion of kilocalories per day of solar load avoided, pounds per year of phosphorus reduced, or the increase in weighted linear feet of salmon habitat now seems obvious, but decades of investments in river restoration have occurred without these metrics—which has left a gap in tracking projects and their achieved outcomes. With the new ability to quantify ecosystem services and uplift, conservation can successfully move forward in a world with increasing demands on limited natural resources (See (Preston 2014) for a discussion of global human development versus ecological sustainability.).

The quantified conservation concept steers organizations such as The Freshwater Trust to further develop and use new metrics, methods and tools to analyze, measure and restore freshwater ecosystems. Work includes scientific research to improve methods of quantifying the ecosystem benefits of restoration actions; technological development to streamline data collection and measurement; and new capacity to collect, store and synthesize data. This research and data management has then been applied by The Freshwater Trust to: edge-of-field sediment runoff models to calculate the reduction in fertilizer flowing into streams from conservation practices on farmlands; a shade model to calculate the temperature benefits of riparian restoration at multiple sites within a watershed; and calibrating a flow augmentation model that calculates the temperature benefits of leaving more water in streams at certain times of the year, used to prioritize landowner recruitment for leasing water rights.

## 7.3  Environmental Markets and Water Quality Trading

Critical environmental challenges to freshwater resources are outpacing the funding available to fix them. At the same time, economic forces are increasingly encouraging conservation funding models that make sense financially as well as environmentally. Therefore, understanding the cumulative negative effect of human impacts on ecosystems—from agriculture to urban development to industrial production—requires new methods and tools, such as market-based approaches, to address those impacts (Horton and Gaddis 2011).

In simplest terms, a market facilitates the trading of goods or services between parties, and "environmental market" means the trading of negative environmental impacts for net positive environmental benefits. Markets can translate the natural functions of a healthy ecosystem into units that can be compared with and traded for impacts—from wastewater treatment, road construction, development, or industrial production—in a standardized and organized way.

Environmental markets are not new. In one form or another they have existed for decades. The carbon market spawned by the Clean Air Act worked well to reduce acid rain in the 1990s and 2000s to 41 % below 1980 levels (EPA 2014). Similarly, the Clean Water Act requirement of "no net loss" to wetlands created wetlands banking—a mitigation market whereby acres of wetlands drained for development are traded for restoration of wetland acres elsewhere. Before the economic downturn slowed the pace of new construction in the early 2000s, Ecosystem Marketplace reported that the wetlands market generated more than $1 billion annually (Madsen et al. 2010). Ecosystem services markets are in many cases becoming the preferred method for conserving ecosystem functioning (Hook and Shandle 2013).

### 7.3.1  Water Quality Trading

As water issues in the U.S. become headline news, there is increased interest in accelerating the restoration of freshwater ecosystems. Water quality trading offers a viable, market-based approach. Under the Clean Water Act, the U.S. Environmental Protection Agency (EPA) and its state government counterparts are empowered to regulate point source entities that discharge into waterways from a pipe—industrial factories, power plants, and wastewater treatment facilities. To determine regulatory compliance, the EPA evaluates each waterway and sets a numeric limit on the maximum load of negative inputs that a waterway can sustain and still meet water health standards. Termed a Total Maximum Daily Load (TMDL), these limits must not be exceeded and the EPA issues National Pollutant Discharge Elimination System (NPDES) permits to entities contributing to the impacts.

For most of the past four decades, regulators have focused their attention on the most visible and urgent impacts—using TMDLs to require point source entities to limit the discharge of toxins, poisons, heavy metals, etc. In aggregate, these entities annually spend billions of dollars to reduce these impacts, such as the City of Portland's $1.6 billion "big pipe" project to limit sewage overflow into the Willamette River. Historically, this money has been spent on expensive technological solutions that address specific water quality parameters, such as improved filtration to remove mercury or copper from a plant discharge.

Cumulatively, these efforts to manage more direct water quality impacts have proven effective, so now regulators are turning their attention to bigger, more broadly distributed problems—namely, water temperature and nutrient overload. For example, water temperatures in many streams in the Pacific Northwest are too warm for fish (EPA 2003a). As a result, agencies that regulate water quality are now requiring facilities to minimize the effect of clean but warm discharge (effluent) entering rivers and streams. In addition, nutrient levels, such as nitrogen and phosphorous, increasingly exceed "drinkable, fishable, swimmable" standards for water quality across the nation, impairing aquatic habitat and contributing to dead zones near estuaries such as in the Gulf of Mexico, Puget Sound and California's Bay Delta. Regulators are actively setting new limits on these nutrients as well.

Water quality trading is a collaborative solution to address water quality and pollutant loads in a watershed, by targeting, calculating and achieving pollutant runoff reductions on (primarily) agricultural lands and trading that benefit to an entity that is regulated for that pollutant. These tradeable units of pollutant reductions are often referred to as credits. Water quality trading thus allows permitted point source dischargers, which have traditionally built engineered projects at their plants, to invest in "green" rather than "gray" infrastructure solutions.

If ecosystems were valued like other marketed products and services, then those communities that need the services that ecosystems provide, such as clean, cool water or runoff control, could be charged a real and sustainable price for those services.

### 7.3.2 The Underlying Science of Environmental Markets

While regulatory effort has driven the mathematics behind measuring environmental impacts (as evidenced by the use of caps on regulated point source entities), until recently there has been little widespread agreement on calculations for quantifying environmental benefits. Without a common denominator that enables comparison of the impacts and benefits, there is little basis for a functioning market.

Consider a key ecosystem service: water temperature. Conservationists have long known that streamside trees provide shade that reduces the temperature of the water by reducing the solar load reaching the water. However, until recently, science did not enumerate the temperature benefits of planting a tree next to a stream—a critical piece of data for trading the temperature impacts of a wastewater treatment plant for the temperature benefits of replanting streamside vegetation.

In 2004 the Oregon non-profit Willamette Partnership worked on the problem of how to measure environmental functions, addressing the key data missing for environmental markets to manifest around water quality. Willamette Partnership's goal was to change how money is spent on watershed restoration. The organization identified water quality trading as a way to move investment from concrete to trees, and the first step was to quantify ecosystem services.

Willamette Partnership pulled together a task force to establish protocols for measuring these services, using calibrated temperature models that had been developed for completing TMDLs in the state. Because temperature limits placed by regulators on point source facilities are also expressed in kilocalories per day, this development allowed the comparison and trade of a facility's kilocalorie debits for kilocalorie credits from planting projects, which could establish an environmental market for temperature.

Beyond developing the package of protocols and quality standards for ecosystem services, the Willamette Partnership also secured agreement from nearly every federal and state natural resource agency in Oregon, as well as a diverse mix of stakeholder groups, on a set of protocols and standards for quantifying benefits of freshwater restoration for application in environmental markets—the first agreements of their kind in the country (Willamette Partnership 2009).

In 2009, the Willamette Partnership completed its Counting on the Environment process, a multiagency agreement on the processes and tools needed to support multiple types of credit trading (e.g., water temperature, salmon habitat, wetland habitat, and prairie habitat). The Counting on the Environment agreement and Oregon DEQ's Internal Management Directive provided the foundation that permit writers, other communities, and third parties could use to robustly explore trading as a compliance option. The vision of Willamette Partnership, The Freshwater Trust and other market proponents is for a multi-service, self-sustaining and replicable marketplace that offers scientifically valid restoration compliance solutions, or offsets, to regulated environmental impacts.

### 7.3.3    Who Is Interested in Markets?

Social impact investors, foundations, government incentive programs, and NGOs that want to show the results of their work and report back to their funders have become interested in environmental markets as a tool to achieve better and more accountable environmental impacts over time.

In conservation, actions that restore ecosystems are generally supported by grants from federal and state agencies and private foundations. Grant programs often fund voluntary conservation actions on private land, using tools and methods largely developed in the 1970s.

Under the current system, grant-seeking entities identify funding sources whose conservation objectives and geographic focus are in line with their project, and then submit a detailed proposal to request support. A typical restoration grant proposal contains three parts. First, a description of the restoration actions to be performed on a project site (e.g., the organization will plant 5000 streamside trees on a 5 acre site).

Second, the cost of the project (e.g., 5000 trees cost $10 each to purchase, plant and maintain for the first few years for a total cost of $50,000). And third, the justification of the project (e.g., the official watershed assessment shows that "a lack of streamside shade due to the removal of trees and vegetation contributes to higher water temperatures that negatively impact native species"). Grant makers evaluate the project proposal alongside many others in a competitive process for limited dollars. Funding decisions are based on a comparison of the relative project merits and cost, the credibility of the grant seeker and the subjective evaluation provided by project reviewers. Then, to assure accountability, grant makers will generally only release funds as project elements are completed. The grant recipient will invoice the grant maker as the trees are purchased and planted, demonstrating that the project was executed as proposed.

Under this system, the grant maker knows that 5000 trees have been planted, but does not know if the project helped to reduce water temperature in the watershed. It is assumed that the trees helped improve stream conditions, but by how much is not known. Further, when projects are evaluated based only on the tasks to be performed, rather than the objective outcome, it is difficult to fairly compare one similar project to another. What if two proposals are submitted, both planting 5000 trees, both costing $50,000, and both on the same river, but only one can be funded?

The evaluation of projects has been subjective for decades because better metrics for evaluating projects have not been available. The emerging discipline of measuring ecosystem services offers better metrics for evaluating and tracking restoration actions, providing new tools to analyze the actual ecological outcomes of regulatory and conservation tracks, and providing valuable information to improve the efficacy of these approaches moving forward.

## 7.3.4   Improving Project Selection

Measuring ecosystem services also allows agencies and organizations to map watersheds in entirely new ways. Combining modern mapping techniques, such as LiDAR (Light Detection and Ranging, an optical remote sensing technology that can measure the properties of a target by illuminating the target with pulses from a laser) with new tools for measuring baseline ecosystem conditions can better target where restoration work needs to happen along a stream. These new maps guide project selection and design.

Returning to the scenario above where two similar proposals are submitted to a grant maker, using quantified conservation outcomes now demonstrate that Project A will reduce temperature loading on a stream at a critical spawning period by 5,000,000 kcal per day and Project B by 10,000,000. Additionally the grant reviewer could have a map that overlays proposed projects with baseline conditions and prioritized sites for shade. Based on the map, Project B delivers greater temperature benefit and is sited within a priority area where the tree planting will maximize shade in places where the fish need it most. These new metrics demonstrate that Project B is better.

Environmental markets and quantified conservation approaches may be used by funding agencies to effectively target grants from federal and state conservation programs by allowing grant reviewers to better consider the planned outcomes of projects. For example, in 2012 the U.S. Forest Service worked with The Freshwater Trust and Willamette Partnership to pilot a riparian restoration program, using a water quality trading program structure. The Freshwater Trust generated credits for temperature benefits via restored riparian buffers at three strategic sites throughout Oregon. These credit-generating projects yielded millions of compliance-grade thermal offset credits, which were verified through Willamette Partnership's Ecosystem Credit Accounting System. The credits were purchased and retired for conservation benefit as part of a grant agreement with the U.S. Forest Service. Further, the funds paid by the Forest Service covered the cost of monitoring and maintaining project sites for 20 years. Other federal, state and private grant makers have expressed interest in the demonstration pilot, as they consider outcome-based approaches within their own grant programs (Horton 2012) (Table 7.4).

## 7.3.5   Impact Investing

Impact investing catalyzes private capital as a supplement to limited philanthropic funding and public-sector dollars. Impact investors approach investing with the intent to generate measurable social and environmental impact alongside a financial return. For example, Equilibrium Capital offers a suite of sustainability funds

**Table 7.4**  Grant funding processes based on tasks or outcomes

| Traditional grant process (Task-based) | |
| --- | --- |
| Proposed project A | Proposed project B |
| Task: Plant 5000 trees | Task: Plant 5000 trees |
| Need: Watershed assessment identifies streamside shade as needed within the watershed | Need: Watershed assessment identifies streamside shade as needed within the watershed |
| Grant requested: $50,000 | Grant requested: $50,000 |
| *Which one should receive funding?* | |
| Grant process using quantified conservation (Outcome-based) | |
| Proposed project A | Proposed project B |
| Outcome: Reduce stream temperature by 5,000,000 kcal per day | Outcome: Reduce stream temperature by 10,000,000 kcal per day |
| Need: Watershed assessment identifies specific kilocalorie reduction targets for the watershed | Need: Watershed assessment identifies specific kilocalorie reduction targets for the watershed |
| Grant requested: $50,000 | Grant requested: $50,000 |
| *Project B receives funding* | |

focused on different sectors. Its Wastewater Opportunity Fund provides capital to construct wastewater treatments plants that not only produce clean, treated water, but also efficiently extract and utilize the energy and nutrients embedded in the wastewater for additional revenue streams.

The more profit-friendly approach has made it easier for major financial service providers to enter the market (Bank 2013). In a short period of time, financial service providers went from screening out of their portfolios holdings with negative impacts, such as tobacco companies, to actively adding in purpose-driven investment opportunities, such as projects creating access to clean water and energy (Bank 2013).

Recently, public-private partnerships between government funders and private investors have formed to explore the "pay for success" (PFS) model for social impact programs. PFS is an innovative financing model that offers new ways for the government to partner with philanthropic and other lenders to provide capital to test promising practices and scale programs that work, significantly enhancing the return on taxpayer investments. PFS maximizes taxpayer dollars by paying for demonstrated results, and allows effective and evidence-based solutions to be identified and implemented (U.S. Government 2013).

PFS programs generally focus on improving the lives of individuals and families and to reduce their reliance on future government services. Programs have addressed energy and water infrastructure upgrades in government-subsidized housing, child health and education, criminal recidivism, and other social services. Governments have also begun addressing climate change and exploring programs to make utilities more resilient in order to reduce potential future impacts from flooding and stormwater damage. So far the focus has been on infrastructure upgrades rather than developing ecosystem-wide resilience.

Johnson notes that scaling conservation programs depends on environmental markets that attract a variety of funders, including profit-motivated investors interested in the "payment for ecosystem services" model (Matzdorf et al. 2014). For a water quality trading program to fit within the PFS framework, program administrators would have to demonstrate that the investment would create both a financial savings for state and local entities and a return on private investor dollars.

## 7.4 Application of Water Quality Trading on the Rogue River

In 2008, the Oregon Department of Environmental Quality (ODEQ) set a TMDL for temperature for the Rogue River, one of the original eight rivers of the Wild and Scenic Rivers Act of 1968. Temperature is important in this region because the waterways support cold-water fish species. The TMDL focuses on the thermal impact of pollution from both point sources and nonpoint sources. Point sources contribute 14 % of the thermal load (warm but clean, treated water), while nonpoint sources contribute 86 % of the thermal load to the river (ODEQ 2006). The new

TMDL included temperature limits on discharge from the City of Medford's wastewater treatment facility on the Rogue.

In 2009, the City of Medford engaged a consulting engineer to evaluate compliance alternatives. One option was a massive refrigeration system to cool effluent as it is discharged from the wastewater treatment facility. While that legally addressed the water temperature load, it impacted other environmental factors through significant power consumption. Another option was construction of a large lagoon to store effluent during the time of year the facility is out of compliance, which could then be pumped into the river at a different time. While legal, the lagoon is an imperfect solution with no real ecological benefit—and both options were expensive to build.

In 2010, Medford and its consulting engineer considered a water quality trading approach as a new alternative to the built solutions. This approach was compared side by side with a chiller and lagoon storage and demonstrated a cost-effective solution with greater environmental benefits. Additionally, the temperature trading protocols developed by Willamette Partnership were approved and promoted by the ODEQ, the regulatory agency that must sign off on the compliance plan (Table 7.5).

In 2011 the City of Medford entered into an agreement to offset the future projected temperature exceedance (the estimated amount that its discharge would be over the temperature limit in a worst-case scenario) through a state-regulator-approved water quality trading program. The offset credit contract directs The Freshwater Trust and its partners to plant and maintain 10–15 miles of streamside vegetation on the Rogue River and its tributaries to reduce the solar load on the water over time and protect critical spawning habitat for salmon.

In return, Medford's Regional Water Reclamation Facility achieves temperature compliance with regulators. This $6.5-million habitat restoration solution proved more cost-effective for the city when compared to the estimated $16-million facility upgrades, and a majority of that investment remains in the community as monitoring and maintenance continues for 20 years (Fig. 7.1).

**Table 7.5** Temperature management alternatives for City of Medford. Present value comparison

| Effluent chillers | $15,140,000 |
| Effluent storage | $15,020,000 |
| Temperature trading | $ 6,500,000[a] |

WYA (2012)
[a]Trading initially priced at $5,130,000 and later revised to $6,500,000

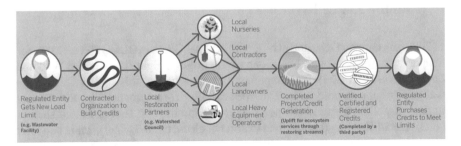

**Fig. 7.1** Restoration model using water quality trading (Source: The Freshwater Trust)

The water quality trading approach functioned as follows:

- A regulated entity contracts with an organization to provide temperature reduction credits—translated as "kilocalories per day" from shade generated by tree plantings—to meet its regulatory temperature requirement.
- Project implementation, credit calculation and landowner relations are all managed by the contracted organization.
- When projects are complete and credits are calculated, a neutral third party certifies the results and the entity is invoiced for the credits.
- The credit transaction includes two parts: a cash payment for the cost of credit generation (the project costs) and payment for the ongoing costs of monitoring and maintaining the sites for 20 years or more—including incentive payments to landowners for acreage converted to conservation use.

The Freshwater Trust packaged all costs into one price per credit that included landowner recruitment, financing of restoration projects, planting, long-term landowner agreements (typically leases), long-term maintenance and monitoring (typically 20 years), and all the verification and reporting needed to keep credits valid. This price covered the same 20-year planning horizon for the facility and was equitably compared to the costs of a built solution.

This model made the restoration alternative as turn-key as the construction of a chiller or lagoon. The restoration alternative was also significantly cheaper than the engineered solutions—roughly half—even with a robust planting regimen and landowner incentives. Further, costs were spread out over time, making it easier to fund than the large, one-time capital expense of a chiller or storage lagoon.

## 7.4.1 Using Data to Improve Selection of Project Types and Locations

The goal of the Medford water quality trading program is to achieve regulatory compliance and reduce solar loading to the Rogue River by blocking 600 million kilocalories per day of solar load by vegetation at maturity. Six hundred million kilocalories per day represents the regulator approved unit of measurement for compliance and is also twice as many credits needed, using a conservative 2:1 trading ratio to account for any uncertainty or force majeure over the life of the program.

Prior to program implementation, The Freshwater Trust identified key areas for restoration using watershed and species planning documents, biological opinions for endangered species, and a variety of remote sensing and GIS (geospatial information system) tools. For temperature concerns, GIS specialists analyzed aerial maps and created catalogs of disturbed and non-disturbed riparian areas alongside rivers and streams. Data included size of disturbed lot, location of lot (south-bank lots provide more optimal shade), ownership, topography, hydrographic data, and location in the watershed (including location near fish passage barriers).

Using its StreamBank® BasinScout™ method, The Freshwater Trust combined satellite photos with physical data about the river, terrain, and land use practices.

**Fig. 7.2** Map of potential riparian restoration sites in the Rogue River basin

The parameters were scored and a map was created with current riparian conditions in the basin categorized and color coded, clearly distinguishing areas in need of restoration from those that were intact and function at a high capacity. Disturbed areas (in red) indicated a lack of shade-producing vegetation near the stream-banks—areas with high potential for environmental uplift. Non-disturbed areas (in green) indicated well-vegetated riparian areas that shaded the streams and thus created favorable habitat for cold-water fish. "Shade-A-Lator", an ODEQ-developed module of HeatSource, was used to estimate the credits by subtracting modeled reductions in thermal loading resulting from riparian restoration at disturbed sites from current thermal loading (Fig. 7.2).

Overlaying layers of data from GIS onto satellite photos outlines precisely how different parcels of land are being used along the river down to the property-owner level. This information is used to assign quantifiable environmental gains that can be translated into credit values. Higher credit values are assigned to sites with the greater potential for environmental improvements.

For example, Little Butte Creek, a tributary to the Rogue River, is one of the best salmon-producing waterways in the basin. Much of the streamside vegetation along Little Butte Creek had been removed to support local agricultural interests, thus reducing shade along the creek. After identifying it as a priority restoration site, The Freshwater Trust planted 4554 native trees and shrubs along 3038 ft of Little Butte Creek as part of the Medford water quality trading program. These restoration actions have calculated benefits to water quality by reducing solar load, limiting nutrient and sediment run-off and stabilizing streambanks. In the long-term, riparian vegetation also serves as a source of large wood and organic matter to the stream that benefits fish and wildlife. Additionally, the landowner at the project site installed a livestock exclusion fence, which further protects the riparian plantings from livestock browse and trampling.

Reductions in solar loading, nutrient and sediment inputs from restoration on Little Butte Creek were quantified by The Freshwater Trust using the Shade-a-lator and Nutrient Tracking Tool (NTT) models. Using pre-project conditions, including the current vegetation height and distribution, Shade-a-lator calculated the pre-project load of solar radiation reaching the surface of the creek. Using mature vegetation height and distribution from planting and project designs, Shade-a-lator predicted the future load of solar radiation reaching Little Butte Creek. The difference between pre-project and post-project solar loading represented the uplift and was reported in kilocalories per day. The Freshwater Trust also calculated nutrient and sediment uplift at the project site using the NTT model. The NTT model considered on-farm drainage patterns, project designs and farm management practices to determine nitrogen, phosphorus and sediment uplift related to riparian restoration and was reported in pounds per year (Tables 7.6 and 7.7).

**Table 7.6** Calculating uplift at Little Butte Creek RM 8.5 using shade and nutrient models

| Solar load avoided | | | |
|---|---|---|---|
| | Kilocalories per day | | |
| Pre-project | 29,947,256 | | |
| Post-project | 8,534,732 | | |
| Uplift | 21,412,533 | | |
| **Nutrients and sediments reduced** | | | |
| | Phosphorus (lbs/year) | Nitrogen (lbs/year) | Sediment (lbs/year) |
| Pre-project | 9.5 | 66.0 | 6989 |
| Post-project | 6.3 | 50.2 | 4412 |
| Uplift | 3.2 | 15.8 | 2577 |

Freshwater Trust (2015)

**Table 7.7** Progress toward generating contracted kilocalories for Medford water quality trading program, 2012–2014

| Site | Planting year | Acreage | Mileage | Phosphorus reduction (lbs/year) | Nitrogen reduction (lbs/year) | Sediment reduction (lbs/year) | Solar load avoided (kcal/day) |
|---|---|---|---|---|---|---|---|
| Rogue RM 128 | 2012 | 3.40 | 0.31 | 0 | 0.5 | 13.6 | 69,073,622 |
| Applegate RM 28.5 | 2013 | 4.70 | 0.56 | 24.4 | 121.1 | 40,117 | 41,809,600 |
| Applegate RM 29.5 | 2013 | 2.60 | 0.30 | 0.4 | 3.8 | 1249 | 23,572,100 |
| Applegate RM 30 | 2013 | 2.40 | 0.31 | 0.4 | 3.7 | 808 | 56,921,925 |
| Applegate RM 3 | 2014 | 6.0 | 0.80 | 8.4 | 98.2 | 39,171 | 56,701,835 |
| Little Butte RM 8.5 | 2014 | 2.73 | 0.58 | 3.2 | 15.8 | 2577 | 21,412,533 |
| Rogue RM 95 | 2014 | 5.92 | 0.77 | 0 | 6.5 | 0 | 67,768,658 |
| **Total** | | **27.75** | **3.63** | **36.8** | **244.6** | **83,935.6** | **337,260,658** |

**Table 7.8** Economic Impacts of Restoration in Oregon

| Project types | Employment (jobs) | Total economic Output ($) | Total wages ($) |
|---|---|---|---|
| In-stream | 14.7 | 2,203,851 | 535,000 |
| Riparian | 23.1 | 2,310,128 | 725,000 |
| Wetland | 17.6 | 2,259,422 | 630,000 |
| Fish passage | 15.2 | 2,240,281 | 573,000 |
| Upland | 15.0 | 2,476,290 | 616,000 |
| Other | 14.7 | 2,270,862 | 560,000 |
| All | 16.3 | 2,311,468 | 589,000 |

Adapted from data in The Economic and Employment Impacts of Forest and Watershed Restoration Nielsen-Pincus and Moseley (2013)

A portfolio of multiple prioritized restoration sites such as Little Butte Creek yielded a comprehensive water quality trading program that is generating the necessary credits for Medford to purchase and achieve regulatory compliance. In the first two years of program implementation (2012-2014), more than 50 % of the credits have been produced, and the program is on track to produce the remaining credits on time and on budget.

In addition to the main environmental benefit of restoring streamside vegetation and reducing solar load, the riparian restoration solution also results in ancillary benefits, including:

- Less energy consumption than gray infrastructure upgrades
- Reduced streambank erosion and silting (Gregory et al. 1991)
- Improved instream habitat for Chinook salmon and steelhead (Gregory et al. 1991)
- Economic impact. Most of the money spent on a restoration project stays in the local economy, employing local nurseries, contractors and other professional services. Every $1 million spent on restoration creates up to 16–20 local jobs (Nielsen-Pincus and Moseley 2013). Restoration projects, as seen in the following table, impact Oregon's economy each year (Table 7.8):

When many water quality trading programs are in place in the most ecologically degraded waterways, they increase the pace and scale of the restoration projects that are essential to addressing the mounting freshwater challenges.

## 7.4.2 Applying the Principles of Water Quality Trading Consistently

While water quality trading has been a discussed concept for more than a decade, it is only recently that all the necessary stakeholders have come together to implement programs in Oregon. The last five years have yielded important lessons for how to move forward with clarity and consistency. Identifying the barriers to trading provides the opportunity to build simple solutions, such as free and available data, prioritized trading areas, and improved guidance for permitting.

Regulated facilities, environmental experts and regulators have worked together to build the science and standards needed to calculate the ecological benefits of restoration actions. These process standards have made it possible for facility managers and permit writers to evaluate the potential of trading as a compliance option for incorporation to NPDES permits and also have made it possible for project developers and engineers to estimate the availability and costs of providing credits.

Examples of standard program elements include (Cochran et al. 2015):

- Using TMDLs to define trading areas and setting eligibility criteria for landowners.
- Establishing site-specific "regulatory baselines" to meet the pre-existing federal, tribal, state, and local regulatory requirements applicable at sites
- Establishing site-specific pre-project conditions to ensure that the credits sold only represent the additional water quality benefit resulting from eligible projects.
- Describing performance standards for restoration projects to make sure they generate ancillary benefits, such as wildlife habitat and floodplain restoration, not just maximize shade.
- Providing for third-party verification and registration of credits.
- Setting guidelines for land protection and maintenance.
- Establishing trading ratios and other risk mitigation measures to cover the time lag in generating shade and associated uncertainty (Table 7.9).

Building on its work from 2004 to 2009, Willamette Partnership and The Freshwater Trust, with water quality agency staff from Idaho, Oregon, and Washington and U.S. EPA Region 10, convened in 2013–2014 and developed joint regional recommendations on improving water quality through trading. These recommendations are meant to guide and assist the agencies and stakeholders in making key decisions when designing and launching water quality trading guidance, frameworks, and plans consistent with the Clean Water Act, TMDLs, and other relevant regulations (Willamette Partnership 2014). In 2015 the National Network on Water Quality Trading took these guidelines to a national scale with a revised publication of the program recommendations that are quickly becoming standardized.

## 7.4.3 Building Monitoring into a Program

Monitoring, maintenance and reporting is an important step to the process. All water quality programs should be required to include long-term monitoring, maintenance and tracking to gauge the ecological benefit of the project. The typical monitoring timeframe is 5–20 years. In the case of temperature benefits, monitoring

**Table 7.9** Temperature credit transaction model for water quality trading

| Project Activity | Description |
|---|---|
| Financing | The project developer secures funds to fully finance credit generation. Regulated entities pay only for registered credits and bear no financial risk until credits are produced. |
| Site prioritization and targeting | After confirming additional weighting factors for site prioritization with local partners and integrating spatial data, the project developer creates a list of sites that meet both thermal credit requirements and address other high-priority ecological needs. |
| Landowner recruitment and site eligibility validation | The site prioritization list is used to contact high-priority landowners. A project developer selects partners as needed, and completes on-site inspections to collect data and ensure eligibility of selected sites. |
| Landowner agreements | Landowner agreements are formed between a project developer and landowners. All agreements run with the land, are recorded in the county land office, name the city as the beneficiary, and are transferable to other parties. |
| Site evaluation, baseline credit determination, and project design | Monitoring points and pre-project data are collected at sites and loaded into a mapping program with other information to complete functional pre-project assessment. Following third-party revegetation standards, local partners create project designs for each site. Designs include materials acquisition, planting layout, maintenance planning, and an irrigation plan. |
| Project planting implementation | Temperature credit projects must meet rigorous implementation and performance standards. The project developer has service contracts with experienced restoration professionals to carry out all site preparation, planting, and maintenance required on the site. |
| Maintenance, monitoring, and reporting | Routine site maintenance begins directly following first plantings. Plantings are irrigated as needed, and invasive plant control is ongoing until "free-to-grow" state is achieved. In-fill planting is undertaken as required. Monitoring is robust during the plant establishment period. Annual on-site inspections and reports to the buyer and regulatory agency confirm that projects are performing to required credit standards. |
| Third-party verification | Project developer continues to work with the designated third party to independently verify that projects are meeting performance standards. Third-party review of monitoring reports likely occurs annually. On-site verification occurs at agreed-upon intervals (thus far, The Freshwater Trust and Willamette Partnership expect on-site verification on approximately 5-year intervals for the life of the credits). |
| Credit registration | Each site is listed on an environmental registry and every credit is assigned a serial number to confirm the validity and custody of credits over time. The project developer generates reports for the buyer and regulatory agency to confirm credit availability for permit compliance. |
| Valid credit certificate and invoicing | Following project implementation, verification, and credit registration, the project developer invoices the buyer and sends a notice of sale to the environmental registry. The confirmed credits are transferred from project developer to buyer for immediate use in permit compliance. |

Cochran et al. (2015)

of riparian trees planted until they have reached a "free to grow" state (approximately 3–7 years) is important to provide assurance of real environmental benefit. In the case of sediment and nutrient benefits, on-farm irrigation practices also require maintenance and monitoring to a set of quality standards, as this long-term commitment can be a significant cultural shift for landowners.

Monitoring for conservation actions can take two approaches. Practice-based monitoring assumes that when a restoration action or management practice is in place, it operates at 100 % efficiency and the modeled outcome will be realized in full. Annual verification procedures assume that the site-specific efficiency rates modeled through an approved quantification tool are realized. On the other hand, performance-based monitoring requires substantial data collection in the field. Annual verification procedures confirm or adjust site-specific modeled efficiency rates. As data are collected, the efficiency would be adjusted to account for new information, but field data collection can be time, labor, and cost intensive. A hybrid approach of practice- and performance-based monitoring is generally recommended to effectively measure the ecological benefits while balancing economic costs of a program.

Evaluating program results over the long-term makes it possible to improve on restoration approaches through adaptive management techniques. Despite its importance, long-term monitoring is traditionally not funded or underfunded with conservation grants. However, environmental markets that focus on outcomes and water quality trading programs that require compliance can allocate the necessary additional funding to ensure that projects are designed to measure long-term results.

## 7.4.4  Water Quality Trading Is Not for Every Watershed

Watershed-wide compliance programs can be appropriate solutions for some, but not all, scenarios. To better understand if a restoration solution is an option, a regulated facility must analyze what role its discharge plays in the impairment of a watershed. If its discharge plays a small role, and if other sources of pollutants play much larger roles, then a restoration solution may work.

Water quality trading is generally supported when it is consistent with the 2003 U.S. EPA Water Quality Trading Policy (EPA 2003b) and where it meets the following criteria (excerpted from (Mascia et al. 2015)):

- **More effectively accomplishes regulatory and environmental goals**
  Water quality trading is supported when it achieves more pollution reduction and greater improvements to water quality than would have occurred without trading over a comparable period of time, and does so with reasonable and predictable costs. Establishing defined trading areas that coincide with a watershed or TMDL boundary results in trades that affect the same water body or stream segment and helps ensure that water quality standards are maintained or achieved throughout the trading area and contiguous waters.

- **Is based on sound science**

  Water quality trading is supported when program goals, credit quantification methods, and adaptive management systems are based on sound science and on their ability to achieve real benefits from an ecological standpoint. Because science evolves, trading frameworks and trading plans should monitor and evaluate outcomes to regularly improve and report on the progress toward water quality goals.

- **Provides sufficient accountability that promised water quality improvements are delivered**

  Water quality trading guidance, frameworks, and plans should seek to foster transparent information on trading rules and processes, location, and volume of transactions, as well as the effectiveness of trading over time. Trading programs should incorporate practices that promote visibility between buyers, sellers and permitting agencies, as well as between the program and interested stakeholders. Transparency can be achieved by directing trading activity through a tracking system and public registry.

- **Does not produce localized water quality problems**

  The use of water quality trading is not supported where it leads to "hot spots" or areas of localized water quality impairment between the point a credit is generated and the point it is actually applied

- **Is consistent with the Clean Water Act regulatory framework**

  Water quality trading should be consistent with the relevant provisions of the Clean Water Act. This includes avoiding trading where it would circumvent the installation of minimum treatment technology required by federal and/or state regulations at the site of a point source, adversely affect water quality at an intake for drinking water supply, delay implementation of a TMDL, or cause the combined point source and nonpoint source loadings to exceed the cap established by a TMDL.

## 7.5   Increasing Pace and Scale

Market-based restoration solutions can have a revolutionary impact on the pace and scale of conservation. The existing systems for restoration in the U.S. rely heavily on grant funding from public and private sources. Although there are billions of dollars available, it is only a fraction of what is needed to advance a nation's worth of environmental improvements.

With the Medford example, one municipality is injecting several million dollars in restoration funding into a single watershed within a five-year period. The majority of the restoration investment will remain in the community. Local landowners who voluntarily allow trees to be planted on their property will benefit from annual payments for use of their land. The payments can catalyze action on the part of landowners who generally do not have the financial resources or willingness to forego agricultural revenue to undertake restoration projects on their land.

Though water quality trades have already occurred in isolated pockets across the country using a variety of approaches, for market advocates, the scenario playing out in Oregon may prove the replicable model needed for markets to work at scale. In addition to the Rogue basin, other municipalities in the state have either committed to or are considering adopting this restoration model to offset their temperature impacts. For the first time, a state-approved system for calculating temperature credits is being broadly applied in multiple watersheds.

Having a model that works at scale is the key to launching market efforts nationwide. With 200,000 regulated entities in the U.S. that have impacts that water quality trading can address, the potential for watershed improvement equals billions of dollars (Whitworth 2015a).

Going forward, markets are part of the solution, but they must work in concert with other conservation strategies. Traditional conservation tactics—preservation, adjusting consumer behavior, point source impact reduction and enforcement—remain key conservation tools. Markets will not and should not replace those efforts. The U.S. Department of Agriculture, a supporter of the water quality trading concept, notes that trading complements existing conservation efforts by providing additional resources for water quality improvement and associated environmental benefits, such as air quality improvements and creating and enhancing wildlife habitat (NNWQT 2015). The USDA also notes that environmental markets must be designed to be self-sustaining over the long run (Horton and Gaddis 2011).

Environmental markets and quantified conservation can provide diagnostic tools and treatments for natural resources, ultimately maximizing the impact of the limited investment in their upkeep. Proper equivalencies or "units of environment" allow tradeoffs that render environmental gain and economic gain—and disallow transactions that do not do both. Market-based, quantifiable methods may help assure that what is spent on the environment actually makes a difference. With pressures on ecosystems mounting, there remains little time for wasted effort.

# References

Bank D (2013) Impact investing 3.0. Freshwater Magazine, vol 5. The Freshwater Trust, Portland
California (2015) State of California. http://ca.gov/drought/
CBF (2014) Chesapeake Bay Foundation. http://www.cbf.org/about-the-bay/issues/polluted-runoff
Cochran B, Primozich D, Swanson K, Wigington T (2015) Advances in water quality trading as a flexible compliance tool (trans: Trading OsCSitEoWQ). Water Environment Federation, Alexandria
Costanza R (1991) Ecological economics: the science and management of sustainability. Colombia University Press, New York
Ehrlich PR, Ehrlich AH (1981) Extinction: the causes and consequences of the disappearance of species. Random House, New York
EPA (2003a) EPA Region 10 guidance for Pacific Northwest State and tribal temperature water quality standards. U.S. Environmental Protection Agency, Region 10 Office of Water, Seattle
EPA (2003b) Water quality trading policy. U.S. Environmental Protection Agency, Washington, DC

EPA (2014) Reducing acid rain. U.S. Environmental Protection Agency, Washington, DC

EPA (2015) U.S. EPA Office of Wetlands, Oceans and Watersheds. http://water.epa.gov/type/watersheds/named/msbasin/zone.cfm

Freshwater Trust (2015) 2014 uplift report. The Freshwater Trust, Portland

Gregory SV, Swanson FJ, McKee WA, Cummins KW (1991) An ecosystem perspective of riparian zones. Bioscience 41(8):540–551

HCN (2014) High Country News. https://www.hcn.org/articles/progress-report-for-colorado-river-experimental-pulse

Hook PW, Shandle ST (2013) Navigating wetland mitigation markets: a study of risks facing entrepreneurs and regulators. Ecosystem Marketplace, Washington, DC

Horton A (2012) Restoration navigation: charting a new course for con-servation investment. Freshwater Magazine, vol 4. The Freshwater Trust, Portland

Horton A, Gaddis M (2011) Pace and scale: how environmental markets could change conservation for good. Freshwater Magazine, vol 3. The Freshwater Trust, Portland

Madsen B, Carroll N, Moore Brands K (2010) State of biodiversity markets: offset and compensation programs worldwide. UNDP, New York

Mascia TJ, Smith B, Furia J, Power K, Wigington T (2015) Advances in water quality trading as a flexible compliance tool. Legal challenges to trading. Water Environment Federation, Alexandria

Matzdorf B, Biedermann C, Meyer C, Nicolaus K, Sattler C, Schomers S (2014) Paying for green? Payments for ecosystem services in practice: successful examples of PES from Germany, the United Kingdom and the United States. Federal Ministry of Education and Research, Muncheberg

Nielsen-Pincus M, Moseley C (2013) The economic and employment impacts of forest and watershed restoration. Restor Ecol 21(2):207–214

NNWQT (2015) Building a water quality trading program: options and considerations. National Network on Water Quality Trading, Portland

NOAA (2015) NOAA, partners predict severe harmful agal bloom for Lake Erie. National Oceanic and Atmospheric Administration. http://www.noaanews.noaa.gov/stories2015/20150709-noaa-partners-predict-severe-harmful-algal-bloom-for-lake-erie.html

ODEQ (2006) Willamette basin total maximum daily load. Oregon Department of Environmental Quality, Portland

Preston M (2014) Sustainability and corporate responsibility: a new language for decision making https://vimeo.com/69700372

Shiklomanov IA (1999) World water resources: modern assessment and outlook for the 21st century (summary of world water resources at the beginning of the 21st century, prepared in the framework of the IHP UNESCO). Federal Service of Russia for Hydrometeorology & Envi-ronment Monitoring, State Hydrological Institute, St. Petersburg

U.S. Government (2013) Strategies to accelerate the testing and adoption of pay for success Pfs financing models, vol 78. U.S. Government Printing Office, Washington, DC

Whitworth J (2015a) Quantified: redefining conservation for the next economy. Island Press, Washington, DC

Whitworth J (2015b) Quantifying conservation for the world ahead. Freshwater Magazine, vol 7. The Freshwater Trust, Portland

Willamette Partnership (2009) Ecosystem credit accounting: pilot general crediting, vol 1.0. Willamette Partnership, Portland

Partnership W (2014) Regional recommendations for the Pacific Northwest on water quality trading. Willamette Partnership and The Freshwater Trust, Portland

WRI (2005) Ecosystems and human well-being: biodiversity synthesis. Millennium ecosystem assessment. World Resources Institute, Washington, DC

West Yost Associates (2012) Plan 2012. West Yost Associates – City of Medford regional water reclamation facility facilities. West Yost Associates, Eugene

# Chapter 8
# Incentivizing Impact – Privately Financed Public Health in Rwanda

**Evan A. Thomas, Christina Barstow, and Thomas Clasen**

**Abstract** There is effectively universal agreement that clean air and clean water are human rights. Yet there is not universal agreement on effective ways of ensuring these rights. In environmental health interventions addressing water and indoor air quality, multiple determinants contribute to adoption. These may include technology selection, technology distribution and education methods, community engagement with behavior change, and duration and magnitude of implementer engagement. In Rwanda, while the country has the fastest annual reduction in child mortality in the world, the population is still exposed to a disease burden associated with environmental health challenges. Rwanda relies both on direct donor funding and coordination of programs managed by international non-profits and health sector businesses working on these challenges. In this chapter, a program in Rwanda illustrates the potential of public-private partnerships, combined with objective measurement tools and metrics, to deliver a sustained impact in poor households.

**Keywords** Rwanda • Water • Cookstoves • Public health • Carbon credits • Sensors

Portions of this chapter are adapted with permission from Barstow et al., "Designing and Piloting a Program to Provide Water Filters and Improved Cookstoves in Rwanda", PLOS One, 2014, and portions adapted with permission from "Use of Remotely Reporting Electronic Sensors for Assessing Use of Water Filters and Cookstoves in Rwanda", *Environmental Science and Technology, 48 (12) DOI:* 10.1021/es503155m. 2014. American Chemical Society. Authors Thomas and Barstow are compensated consultants to the implementer described, DelAgua Health. Authors Thomas and Barstow are responsible for this chapter with the exception of section 8.9, written by co-author Clasen.

E.A. Thomas (✉)
Department of Mechanical and Materials Engineering, Portland State University,
1930 SW 4th Ave, Portland, OR 97201, USA
e-mail: evthomas@pdx.edu

C. Barstow
Civil, Environmental and Architectural Engineering Department, The University
of Colorado at Boulder, UCB 428, Boulder, CO 80309-0428, USA

T. Clasen
Rollins School of Public Health, Emory University, 1518 Clifton Rd, Atlanta, GA 30322, USA

© Springer International Publishing Switzerland 2016
E.A. Thomas (ed.), *Broken Pumps and Promises*,
DOI 10.1007/978-3-319-28643-3_8

## 8.1 Background

Access to improved drinking water and improved stoves could benefit the millions who suffer from diarrheal disease and pneumonia, two of the leading causes of death around the world for children under five. Worldwide, of the 7.6 million deaths in children under 5 in 2010, 64 % were associated with infectious diseases including 18 % with pneumonia and 11 % with diarrhea. Combined, pneumonia and diarrhea kill over 2 million children each year (Liu et al. 2012).

Some of these deaths may be avoided through interventions to improve indoor air quality and household water quality: pneumonia is often linked to indoor air pollution from biomass fuels (Smith et al. 2000; Ezzati and Kammen 2001) and diarrhea to deficiencies in water and sanitation, including poor water quality (Black et al. 2003). Many cookstove interventions have shown a reduction in indoor air pollutants such as carbon monoxide and fine particulate matter (Roden et al. 2009; Masera et al. 2007). Similarly, interventions targeted at improving household water quality through the implementation of water treatment strategies such as chemical treatment, boiling, solar disinfection or filtration have been shown to reduce diarrheal disease (Clasen et al. 2007; Fewtrell et al. 2005).

The national environmental health program described here evolved from experience authors Thomas and Barstow and our team have had designing programs in Rwanda since 2003. These first interventions, run as Engineers Without Borders-USA, included small-scale water treatment, rainwater catchment and cookstove technologies. Over time, it became apparent that grant based programs have difficulty scaling and maintaining performance without significant and continuous funding inputs.

An introduction in 2007 to the United Nations Clean Development Mechanism 'carbon credit' system created under the Kyoto Protocol inspired a funding model that could earn, and sell, carbon credits associated with objective evidence of drinking water treatment, thereby generating revenue directly aligned with the intent and investing much of that revenue in sustaining these interventions.

We were the first to register any United Nations program for drinking water treatment, piloting the concept in Rwanda. We contracted our expertise to a water filter manufacturer, Vestergaard Frandsen, assisting in the development and deployment of a 4.5 million person program in Kenya in 2011. In 2012, we were contracted, and later acquired, by DelAgua, a water product supply company, to scale our program in Rwanda using private finance.

Public and heated debates about the Kenya program included suspicion of private enterprise in this type of public health work; skepticism regarding the carbon credit model; and highly contradictory reporting of water filter use by a variety of organizations.

To address some of these concerns, our Rwanda program includes an independently managed health impact randomized controlled trial, led by author Clasen, and leverages cellular network reporting sensors installed in a sample of the stoves and water filters to directly measure use. These controls help align the intent of the program – a health impact – with the revenue earned from the credits. This chapter presents several facets of this effort, which is a work in progress.

## 8.2 Rwanda Today

Small, landlocked Rwanda is among the United Nations Least Developed Countries, with a population of nearly 12 million (CIA 2012), at least 80 % of whom live in rural areas with a GDP per capita of about 500 dollars, nearly 100 times less than the United States (Stats 2012). Yet, Rwanda has managed to accomplish something the United States has not – nearly universal health care coverage with primary care educators working in every village across the country. These efforts have resulted in the fastest annual reduction in child mortality in the world, 11.1 % annually between 2000 and 2011 (UNICEF 2015), and contribute to Rwanda now having one of the fastest growing economies in Africa (IMF 2012).

### 8.2.1 Rwanda Public Health

The genocide in Rwanda in 1994 killed over one million people and dislocated two million more in a country that only had about nine million people. Beyond the deaths and disruption, there was extensive destruction of social services, including Rwanda's health care system both through the death and dislocation of professionals, and the destruction of equipment, clinics and hospitals, as well as a broader breach of faith in government and social support.

However, since that time, the government has taken extraordinary measures to improve the quality and performance of health services, from the national level down to village directed efforts. Today, Rwanda has the fastest rate of child mortality reduction in the world (UNICEF 2015).

The Rwanda Ministry of Health (MOH) is the focal point for government health policy, as well as the coordinating entity for international donor efforts.

As a developing country, Rwanda is not yet able to finance all health service activities directly, and relies both on direct donor funding to government programs, as well as careful coordination of programs managed by international non-profits and health sector businesses.

The Rwandan Community Based Health Insurance Policy identifies poverty and health as interlinked, including noting that poor health can lead to poverty and that poverty is often the root of many health problems. Therefore, a primary health mission is to decrease poverty in Rwanda, and improve financial accessibility to health services. One of the main methods is through "strengthening community participation and ownership" (Rwanda-MOH 2010).

This is achieved primarily through the Community Health Worker (CHW) program. The CHW system includes three (with plans to expand to five) CHWs per village of 100–150 households (about 500–900 people). The system hopes to target 80 % of health issues in Rwanda (Health 2012). The CHWs are part-time employees of the sector heath centers, and provide basic services such as maternal and newborn health monitoring, vaccination advocacy, family planning, and sanitation and hygiene education. The latest MOH strategic plan targets reductions in acute respiratory infections, malaria, diarrhea, HIV and malnutrition in children, and will add in mental health services, and are primary executed through the CHW program. This program shares credit with other initiatives for the impressive reduction in these diseases annually in Rwanda (Rwanda-MOH 2012).

Other development partnerships with the government have successfully used the CHW structure for programmatic implementation, including for early childhood development education and prevention of HV transmission from mother to child (Lim et al. 2010).

The CHWs are also responsible for administering the national health insurance *Mutuelle de Sante*, including collecting premiums (Saksena et al. 2011). These premiums are charged on a sliding scale according to ability to pay, and are tied to a village-level evaluation of socio-economic condition of each household called the *Ubudehe* process.

The Ministry of Local Government administers the Rwandan *Ubudehe* program, which empowers local villages, called *Umudugudu*, to identify community priorities and work collectively to address them. These community priorities are integrated into cell and sector level development plans.

Each *Umugugudu* works through the *Ubudehė* structure to identify the economic condition of households in their community. Each household is placed into one of six *Ubudehe* categories. These range from Category 1, the poorest, often defined as households without land and that face difficulties finding food, to Category 6, land holders with relatively significant material assets, education and employment. Data collected by the European Commission in 2009 suggests that the program has been highly effective at increasing the welfare of communities. 95 % of households reported that incomes had improved since the program was established, and 89 % responded that the program had resulted in improved social cohesion. While the *Ubudehe* program may well not be responsible for all the improvement in wealth, it is likely a contributing factor. The program was recently awarded the United Nations Public Service Award (Niringiye and Ayebale 2012).

The poorest two economic levels, *Ubudehe* Categories 1 and 2 are defined on the *Umudugudu* (village) level, and that census, collected yearly, is transmitted to the local clinics, district hospitals, and national MOH. This empowers each village to define for themselves who are too poor to pay for health insurance, and thereby receive it for free from the government (Health 2012).

## 8.2.2 Drinking Water

While Rwanda has demonstrated significant progress towards the Millennium Development Goals, almost 30 % of households do not have access to an improved water source (Rwanda 2011), and the improved water sources may become contaminated during collection, transport or storage within the home (Gundry et al. 2004; Onda et al. 2012). Once water is in the households, less than half (46.1 %) of rural

Rwandan families report treating their drinking water, with boiling as the leading treatment method (38.1 %) (Rwanda 2011), which again can become recontaminated after treatment (Wright et al. 2004).

46.1 % of rural Rwandan families take some measure to treat their water, with the leading treatment being boiling (38.1 % followed by chlorination (13.7 %), leaving 53.9 % of people doing nothing to treat their water (Republic of Rwanda 2011)). About 28 % of rural Rwandan households do not have an improved sanitation facility and only 10 % of households have a designated place for handwashing. Meanwhile, the prevalence of diarrhea (defined as diarrhea in the 2 weeks preceding the survey) was 12.7 % among "improved" water source users and 14.5 % among "not improved" (Rwanda 2011).

The Rwanda Standard for Potable Water defines the microbiological limit of potable water to zero coliform forming units (CFU) per 100 ml of total coliforms. A baseline water quality assessment of 230 improved water sources, 78 unimproved water sources, and stored water in 468 households across all 30 districts in Rwanda, indicated that 27.8 % of improved water sources, 80.2 % of unimproved water sources, and 58.3 % of stored household water supplies exceeded this standard (Kirby 2013), falling into the "intermediate", "high" or "very high" risk World Health Organization (WHO) categories (WHO 1997) for biological contamination of drinking water supplies. Another study of households within the other 11 districts of the project area prior to the start of the program implementation indicated that 81.1 % of households exceed this standard, with 59.1 % falling into the "intermediate" or "high" risk categories (Rosa 2012).

## 8.2.3  Household Energy

About 2.7 billion people use biomass on open fires every day for their daily energy needs, and two million people die each year from illnesses including pneumonia and other respiratory infections associated with indoor air pollution. Of the pneumonia deaths in children under five, about half of them are due to indoor air pollution. The use of biomass fuels contributes to deforestation, climate change, and is generally a time consuming and low productivity fuel source (WHO 2010). In particular, the use of three-stone fires indoors with woodfuel leads to high levels of indoor air pollution. The WHO is currently working on standards for indoor air pollution, but they do not yet exist to the same level as the comparable design and testing standards for household water treatment.

In Rwanda, only 10 % of households have electricity, and 83.3 % of rural households use wood fuel for their daily energy needs (Rwanda 2011), in a country that the United Nations states uses woodfuel at a rate that is 98 % non-renewable (UNCDM 2013). With respect to environmental health associated with the use of wood fuel, the Rwanda HSSP highlights a gap in the existing programmatic activities to address indoor air pollution (Health 2012).

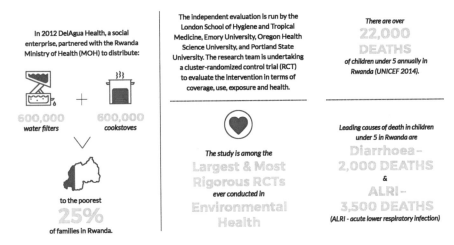

**Fig. 8.1** DelAgua Health and Rwanda Ministry of Health Tubeho Neza Program outline

## 8.3    Tubeho Neza – Let Us Live Well

DelAgua Health, a for-profit social enterprise, was established to combine household technologies that address environmental health issues with market-based mechanisms. DelAgua Health participates in the United Nations Clean Development Mechanism (CDM) to earn carbon credits associated with the reduced use of, and demand for, fuel wood associated with water treatment and cooking, and then sell those credits to buyers as a way to recover costs and profit (Thomas 2012).

In Rwanda, DelAgua Health is partnered with the Ministry of Health since 2012 with a goal to distribute free of charge household water treatment and high efficiency cookstoves to approximately 600,000 households (about 3 million people), throughout the country's 30 districts. The project targets *Ubudehe* categories 1 and 2, the government-recognized poorest 25 % of the country (Fig. 8.1). As of publication, the program has reached over 460,000 people with water filters and cookstoves, and is on track to reach a further million people with cookstoves by the end of 2016.

### 8.3.1    The Products

The water filters used in this program, the Vestergaard Frandsen LifeStraw Family 2.0 is a point-of-use microbial water treatment system intended for routine use in low-income settings. The system filters up to 18,000 liters of water (Clasen et al. 2009), enough to supply a family of five with microbiologically clean drinking water for at least two years, thus removing the need for frequent (weekly or monthly) intervention. The system exceeds the 'highly protective' WHO Standard for household water treatment technologies (WHO 2011; Narajo and Gerba 2011). In a recent study, this filter was shown to be highly effective in improving water quality and was protective against diarrhea, reducing incidence by over 50 % (Peletz et al. 2012).

The cookstove used in this program, the EcoZoom Dura, is based on the rocket-stove concept that is designed to concentrate the combustion process while channeling air flow to create a more complete burn. A complete burn of carbon rich material will also result in little to no smoke. This complete burn utilizes all combustible material creating intense heat and leaving small amounts of residual material. In the field, performance varies but a properly used rocket stove will significantly reduce woodfuel use by about half, although reductions in indoor air pollution vary between designs, fuel types and use. The thermal efficiency of this stoves is 38 % (ARC 2012).

## 8.3.2   The Methods

The primary purpose of the technologies provided is to realize a health benefit. As a first step, communicating these potential health benefits to a user is often seen as an appropriate prerequisite to adoption. A lack of knowledge of potential health benefits has been shown to result in poor adoption of products like stoves and filters (Aboud and Singla 2012). However, it has also been demonstrated that knowledge of health benefits alone is not sufficient to result in sustained behavior change in an individual or household (Rosenstock et al. 1988). The program studied here uses both health based messaging as well as economic and social messaging to promote behavior change within the program.

The program is designed leveraging established behavior change theories, including the Diffusion of Innovation theory (Rogers 2003) and the Health Belief Model (HBM) (Becker 1974). In particular, the program design assumes that continued, comprehensive engagement is critical in order to effect positive behavior change. The program design takes a hybrid approach, integrating pieces of these theories that apply in a Rwandan context to shape the communication strategy of the program.

Several components of the Diffusion of Innovation theory are applied to the program. At the initial distribution meeting, community members are informed about the potential health and other benefits of the water filter and cookstove creating 'initial knowledge' around the technologies. The 'persuasion stage' is initiated through demonstrations at both the community meeting and the household. The stove demonstration includes assembling the stove, how to adjust a pot skirt which can be fitted to different sized pots and finally a fire being started in the stove with some demonstrations including boiling a pot of water to show the rapidity of the cooking process. The water filter includes a demonstration of filtering visually dirty water with clear water coming out of the tap and the maintenance procedure, which includes backwashing the filter. Progressing to the 'decision stage' the household is then asked to demonstrate use and maintenance of the technology, allowing them to trial the technology. Households then move into the 'implementation stage' where they can choose to adopt or reject the technology. About a month later, the program implements the 'confirmation stage' where households who have chosen to partially adopt or reject the technology are given additional training and messaging to hopefully reverse their decision.

Through this program CHWs play several important roles including informing households of the need for the devices, encouraging adoption and analyzing potential problems with the technologies. The CHWs play an especially important role with the 'late adopters' and 'laggards' as more effort is needed to change the household's old habit and promote the new behaviors (Aboud and Singla 2012).

The Health Belief Model is used to shape messaging. The belief that there is a health threat is compelled by messaging related to clean drinking water and clean indoor air. Households are educated about the reduced risk of diarrheal disease from water borne diseases and the reduced risk of respiratory problems from breathing indoor air pollutants. Additionally an important concept in social cognitive theory is often added to the health belief model, self-efficacy, which states that the user must believe that they can adopt the new behavior (Rosenstock et al. 1988). This is facilitated through households gaining confidence in the use of the technologies by having members of the household demonstrate proper use.

While the HBM provides relevant guidance on behavior change theory it is important that the program also express non-health benefits to users. Previous interventions related to both water quality and improved cookstoves emphasize the need to highlight non-health benefits such as those related to economic and social benefits (Aboud and Singla 2012; Figueroa and Kinaid 2010). Thus CHWs educate households on additional benefits such as reduction in medical costs from the water filter and a reduction in cooking time and expenditure on fuel costs for the cookstove.

### 8.3.3   Phase One

Before the execution of the *Tubeho Neza* program, a two year Phase One program of 15 villages and approximately 2,000 households was conducted to assess feasibility and provide insight into the logistical and educational components of the program.

During the two year Phase One process, educational household visits were performed by CHWs monthly for the first 5 months following the distribution of the technologies, again 9 months after distribution, and then 18 months after distribution before being integrated into the larger Phase Two program after approximately two years. Household visits were used to monitor on-going adoption and use of the technologies while piloting and refining educational messaging for the larger program.

Additional activities included a focus group with CHWs approximately three months after distribution to assess CHWs perceived usage of the technologies within their communities and piloting of promotional education activities at several community level events including village meetings, market days, a monthly meeting of parents to discuss topics related to children's health, and the monthly mandatory community service day (*Umuganda*). Additionally, randomly selected unannounced household visits to further assess technology adoption were conducted in approximately a dozen households in each village nearly one year after technologies were in households and at 18 months post distribution.

The Phase One program yielded several lessons which were then integrated into the large scale Phase Two program.

## 8.3.4  Phase Two

Phase Two was deployed in late 2014 to over 100,000 households in the Western Province of Rwanda (Fig. 8.2). Prior to distributions, 360 unique distribution points were identified, including government offices, schools, churches and health facilities. In addition, an extensive process of updating the beneficiary list was completed before distributions. The list identifying the *Ubudehe* category for each household was completed in 2012, two years before the program. These lists were distributed to village chiefs who were asked to update them based on the current residents of his/her village. After all storage and distribution points and the schedule were established, the Rwanda National Police were responsible for transporting products from the capital to the 360 established locations.

Distributions occurred throughout the Western Province, starting with four distributions in the first week and reaching 59 distributions at the peak of the campaign. On average 31 distributions were conducted per week during the 13 weeks of the campaign. Distributions occurred at the cell level, which on average consists of seven villages. The size of a distribution varied from 25 to 753 households with an average of 256 households per distribution. Given the varying size of a particular cell, distributions took anywhere from several hours to two days.

Each distribution was facilitated by local officials who gave opening remarks regarding the program. CHWs then performed a skit that portrayed a family before and after receiving the water filter and cookstove. The skit ends with the singing and dancing of the "*Tubeho Neza* song" developed for the program. After the skit, households were asked to queue in order to receive the products. Discrepancies or disagreements on distribution lists were arbitrated by village chiefs or local CHW leaders. A separate smartphone-based distribution form was collected for each household which included household identification information, photos and signature of recipients, and barcode scanning of the water filter and cookstove.

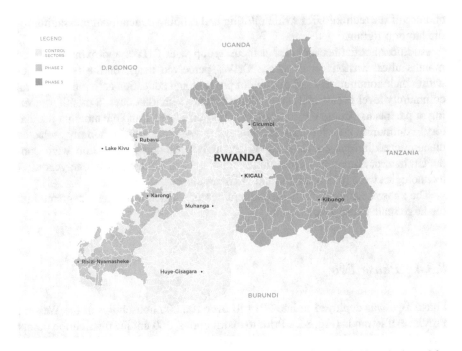

**Fig. 8.2** Tubeho Neza program coverage area. Phase 2 completed in 2014, Phase 3 planned for 2016

Households were then instructed to bring their products home and wait to be visited by a CHW.

Distributions were tracked through electronic forms sent to the DelAgua server after all products were distributed at each site. An online dashboard tracked distribution numbers against expected household numbers from the *Ubudehe* list. Additional analysis was completed on a dashboard to identify duplicate or abnormal forms, which could then be relayed to field staff for arbitration.

### 8.3.5    Initial Household Visit

Following distribution activities, CHWs convened with their DelAgua supervisor to divide up household clusters and visiting routes, devising a strategy for completing all household visits, with input and sometimes accompaniment from authorities most familiar with the particular areas. Rwanda's challenging terrain often meant CHWs had to travel distances of several kilometers to reach beneficiary households. Household visits were performed for a total of 98 days with an average of 1037 household visits performed each day. On average 79 CHWs were performing household visits 6 days a week. At the peak of the program 309 CHWs performed 2274 household visits in a single day. Visits were tracked through a smart phone based form, which could be tracked cross-referencing several parallel identifiers in the distribution forms to determine any households who received products at distribution but had not yet been visited by a CHW. As with the distribution forms, additional analysis was performed to identify duplicate household visits or other possible data entry errors.

Household visits included two components – a brief baseline survey and an extensive education and training session. The survey included baseline fuel, stove, cooking location, water source, and any water treatment methods currently used by the household. Additionally general household identifying information was collected (names, phone numbers, identification numbers, GPS coordinates) and product barcodes of the newly received filter and cookstove were scanned to track products to specific household locations.

Household education included use of interactive teaching tools, primarily an illustration based flipbook and a poster, customized to the household's size and daily routines, which was hung in each household. The design of the flipbook included colorful graphic images illustrated from photographs (example pages shown in Fig. 8.3). Images were piloted with several families to develop appealing and culturally appropriate images. Each page of the flipbook included a specific message to be communicated to the family by the CHW. Instructional pages included a step-by-step process to perform usage and maintenance tasks, while

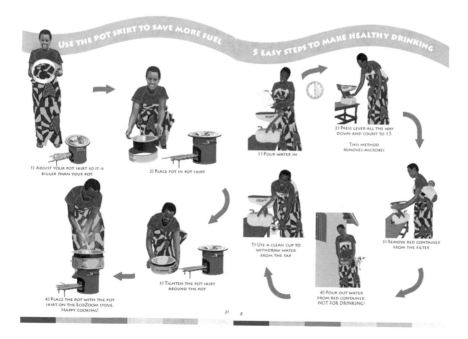

**Fig. 8.3** Example pages from educational flipbook used by CHWs during household education visits

prompting the CHW to physically perform the tasks and have members of the family demonstrate usage. Households had been advised during the distribution meeting to fill the water filter in preparation for the CHW visit, as the initial filling of the backwashing chamber might in some cases exceed household visit time, so that this maintenance feature could be demonstrated with full functionality. The poster included several activities personalized for each family such as circling the number of times to fill a filter in order to provide the entire recommended water consumption amount to all members of a family per day based on its size.

Key messages included:

- Family Oriented – Both the flipbook and poster emphasized ownership of the products by all members, aspiring to be a healthy and happy *Tubeho Neza* family. CHWs were encouraged to engage all available members of a household in the visit.
- Health, Environment and Livelihood Consequences and Benefits – Common diseases and health effects from contaminated drinking water and indoor air pollution were highlighted, as well as possible environmental effects from deforestation. Many of these consequences were then discussed in relation to benefits from using the technologies including financial savings, time savings and cleanliness.
- Comprehensive Filter Description – Phase One households expressed interest in understanding exactly how the filter worked as it was seen as intimidating which made some households hesitant to use and adopt the product. A pictorial description of membrane filtration and the cleaning process was added which helped households understand the importance of backwashing.
- Hydration – In response to skepticism from Phase One households over the program's messaging of the importance of consumption of two liters of water per person per day, messaging was developed to promote hydration through explanation of its health benefits, including reinforcement of the biological importance of water for all ages, young and old.
- Exclusive Use of Filtered Water – Targeted messaging was developed to encourage families to bring filtered water with them to school, work or leisure activities. Families were also asked to designate clean containers as *Tubeho Neza* containers to be used only for safe water storage. A hatch mark was drawn on the containers to distinguish these from others, and households were trained to clean such containers once a week.
- Wood Storage – The difficulty in using the EcoZoom stove with wet or damp wood was indicated by many Phase One households. Households were asked to designate a specific area where fuelwood could be stored so that it could be dry for future use.
- Stove Stacking Behavior – To combat stove stacking (use of traditional stove alongside improved stove), examples of reduced cooking times and fuel consumption from using the improved cookstove were emphasized in the flipbook as well as negative messaging around the use of the traditional three stone fires being harmful and wasteful.

- Cooking Location – To provide additional health benefits related to the use of the improved cookstove, households were instructed to cook outside. However, this was difficult for many households with the large amount of precipitation in much of the Western Province. Therefore, messaging highlighted the portability of the stove, to show it could be moved to a doorway or other household location both well-ventilated and covered.

When CHWs finished the education lesson, beneficiaries were asked to countersign an agreement between their household and a local official acknowledging that the products are for the benefit of the family and are not for sale. A record of this agreement was kept by photographing it using the smart phone. Additionally a cellular phone network shortcode for a repair and replacement hotline was displayed on the poster, which families can contact in case of any problems with the products.

## 8.3.6  Follow Up Household Visits

Following the Phase Two distribution in 2014, a follow up campaign was implemented which consisted of household visits to all households who originally received products. The follow up visits were conducted in spring 2015 between six weeks to six months after households received products. All CHWs were deployed within a five-week time period. On average CHWs performed 1176 household visits per day with a peak of 3557 households visited by 604 CHWs in a single day.

A follow up household visit included a brief survey to assess several adoption and programmatic metrics, repair and replacement of broken products, cleaning of the filter's bottom safe storage water container and an education and training lesson.

The CHW follow up survey questions were focused on current water treatment and cooking practices, primarily assessing initial adoption and continued use of the filter and cookstove. Questions included asking households to report their current household behaviors but additional observational measures were included such as the presence of water in a filter or visible cooking practices occurring during the visit to provide more objective data points. The survey was of similar length to the baseline survey and could be administered in approximately ten minutes.

The household education included emphasis of critical messaging as described previously through similar picture based images presented through a new education material, a yearly calendar, with messaging resembling that used in the original flipbook and poster. Prominence on the calendar was accorded to specific messaging components based on relative priorities of re-visiting, taken from an analysis of the previously mentioned assessment surveys conducted in the quality control activities of the initial household visit campaign. Households were encouraged to use the calendar for their daily lives, as well as events related to the technologies, such as weekly or monthly cleaning tasks. Household members present at the time of visit were again asked to demonstrate use of the products, and CHWs ensured they were able to perform all necessary tasks. Additionally, a *Tubeho Neza* designated safe water sticker, with an illustration of the model *Tubeho Neza* family, was added to safe storage devices previously designated with the *Tubeho Neza* hatch symbol. This was intended, not only to reinforce sanitation behaviors associated with the filter, but also to encourage pride in households' self-identification with the *Tubeho Neza* program when using the safe storage devices out in the community. Finally, all households deemed by the CHW to be correctly using and maintaining the technologies were given a plastic *Tubeho Neza* bracelet as a token of further identification with the program.

To track the follow up campaign, supervisors used a comprehensive smart phone reporting system. Any household that could not be found was reported for as missing, moved, or otherwise unavailable, by supervisors while any unaccounted for product was reported as stolen, sold or at another location. Any product that could not be repaired by CHW's at the time of the visit was reported by the CHW as "in need of repair" in a section in the survey. DelAgua supervisors provided CHW teams with certain filter replacement parts, including taps, backwashing tubes, backwashing container, and pre-filters, to be used in CHW-repairs, which were also tracked through the Follow Up Survey. Additionally, in households found to have sold or attempted to sell one or both of the products, or in households found to have been initially distributed, by mistake, product to which they were not entitled, any remaining product was repossessed by a Supervisor, returned to a local storage facility, and reported.

To combat potential algae growth in the bottom container of the filter, as seen in some Phase One households, a mandatory cleaning of each of the filters was performed by CHWs.

### 8.3.7 Verification Surveys

Two types of survey data are described throughout this chapter; those collected by CHWs from nearly all of all households and data from a verification survey administered to a sample of the households. In this chapter, these surveys will be referred to as "CHW" or "VS" respectively to distinguish the origin of the data. Two rounds of detailed surveys were conducted; one between January and April of 2015, approximately 6 weeks to 6 months after distribution, and the second between July and September of 2015, approximately 10 months to a year after distribution. The verification survey was administered by DelAgua staff and was inclusive of parameters required by the United Nations Clean Development Mechanism monitoring guidelines and methodologies for carbon credits. The CDM also requires a third party auditor to verify survey data and perform field visits to a sample of surveyed households. Additional guidance included a World Health Organization manual on monitoring and evaluation for household water treatment programs (Lantagne et al. 2012). Additional questions were included to assess environmental, health, social and livelihood benefits of the program. The survey consisted of over 100 questions and took approximately 45 min to an hour to administer. The survey was piloted extensively in pilot households and enumerators were required to attend a three day training on administration of the survey including field based practice surveys in households.

The sampling strategy differed in each survey round. For the first verification survey, a two-stage, cluster sample design was employed. In the first stage, 320 villages in Western Province were randomly selected with probability proportionate to size (PPS) sampling (the number of recipient households was used as the measure of size). In the second stage, three households within each village were randomly selected using simple random sampling (SRS). This resulted in a self-weighted sample of 960 households. At the end of the sampling period, an additional 40 households were selected using SRS and added to the sample to meet CDM requirements, bringing the total number of surveyed households to 1000. During the second verification, only a simple random sample was used, for 187 valid surveys. Household that could not be found, did not consent or did not have an adult over the age of 18 responding were not surveyed and the next household in the randomly generated list was visited. To avoid a potential source of survey bias, surveyors were not provided with this list in advance, and instead contacted the survey manager for the next house on the list when necessary.

## 8.3.8   On-Going Promotion Activities

Behavior change and reinforcement activities are ongoing throughout the intervention area. DelAgua staff reside full time in each of the seven Districts of Western Province to manage these activities. Ongoing behavior change activities include:

- CHW Cooperative Meetings – Staff provide additional educational messaging, receive updates on adoption within households and facilitate incorporation of the *Tubeho Neza* program into other health programs.
- Community Meetings – Staff carry out informational sessions which address specific educational goals at common community meetings such as the community service day (*Umuganda*), market days, and other official meetings, as well as to provide repair and replacement services.
- Field and Household Visits – Staff have frequent presence at the household level, through both announced and unannounced household visits to assess technology adoption involving local officials and other local stakeholders.
- Community Hygiene Clubs – Organized activities address community hygiene clubs specifically with benefits and ask members to advocate *Tubeho Neza* products.

DelAgua staff are also responsible for repair and replacement of technologies. Reporting of broken products initiated by households or community leaders calling staff directly or the DelAgua 'shortcode' hotline. Each report is documented and assigned. Staff is then responsible for performing community based repairs or replacements in areas where they are needed, which are reported when completed and tracked in a Work Order system. Replacement parts are stored at the District and Sector level to provide easy access for staff.

### 8.3.9 Product Delivery

A total of 457,778 people across 101,778 households received water filters and cookstoves during the initial campaign distribution. Of these households, 88 % (89,609) were households classified as *Ubudehe* 1 or 2 with the remaining 12 % (12,157) consisting of households from local cell and village officials, local community health workers and pilot households outside of the *Ubudehe* 1 and 2 classification. Following the distribution, community health workers visited 97.8 % (99,515) of households to perform household level education and training activities. Average household size was 4.5 with 0.61 children under five. Before receiving the water filter and cookstove, 89.0 % households reported firewood as their primary fuel source with three quarters (76.1 %) of households reporting the traditional three stone fire as their primary cookstove and the majority (59.2 %) reporting primarily cooking indoors. Most households reported the public tap (43.6 %) or protected spring (31.1 %) as their primary water source with a quarter (26.6 %) reporting treating their water before receiving the filter mostly by boiling (80.7 % of households reporting treating their water).

Overall, 90 % of households identified on the *Ubudehe* list received products. Most households not reached on the *Ubudehe* list were attributed to discrepancies such as households listed multiple times or households which had moved out of the intervention area. Over the course of the initial campaign, 212 (0.2 %) products were repossessed for reasons including allocation to the incorrect household (119, 0.1 %), a household receiving multiple products (59, 0.1 %) or a household selling their filter or cookstove (17, 0.02 %).

The follow up campaign reached 98,804 (97.1 %) of the households which were originally distributed technologies. CHWs recorded just over 1 % of stoves missing (1164, 1.2 %) and under 1 % of filters missing (930, 0.9 %) during the follow up household visits. Missing products were primarily attributed to stolen products (335 (0.3 %) stoves, 138 (0.1 %) filters), sold products (315 (0.3 %) stoves, 261 (0.3 %) filters), products being kept at a relative or neighbor's house (263 (0.3 %) stoves, 208 (0.2 %) filters) and products being stored in a locked room where the CHW could not confirm the presence of the products at the time of the visit (210 (0.2 %) stoves, 254 (0.3 %) filters). Only minor hardware issues with the stoves were reported by CHWs which did not require replacement or repair.

Since the follow up campaign, DelAgua staff have continued to perform repair and replacement activities throughout the intervention area. Approximately 12 months following the original distribution, stoves required minimal maintenance. Filters have required more attention with 187 (0.2 %) filter replacements primarily from households trying to disassemble the filters with staff finding either the water nozzle (83, 0.1 %) or plastic joint connecting the dirty water and safe storage sides of the filter (36, 0.04 %) broken. Additionally 931 (0.9 %) filter repairs have been performed, mostly attributable to the replacement of the backwashing tube (649, 0.7 %) which is more vulnerable to damage because it is the only exposed soft-goods portion of the filter. Other filter repairs included backwashing clogged filters (117, 0.1 %) and the reassembling of the joint between the filter (26, 0.03 %) when it did not require a full replacement.

## 8.4   Water Filter Adoption Indicators

The CHW follow up survey of the majority of households and the two more comprehensive verification survey rounds of a subset of the households, all measured the reported filter adoption above 90 % and observed filter adoption above 75 %. During the CHW follow up visits, 94.1 % of households confirmed treating the last water they consumed with 99.5 % of those households reporting using the LifeStraw filter as the water treatment method (93.6 % filter adoption population-wide). The first verification survey conducted concurrently with the CHW follow up survey, measured 95.9 % treating the last water and again 99.5 % reporting the filter as the treatment method (95.4 % filter adoption including non-treaters). The second verification, performed at least 10 months after distribution showed a small decrease in adoption with 92.0 % of households reporting treating the last water they consumed and 99.4 % reporting the filter as the treatment method (91.4 % filter adoption including non-treaters). Observed filter adoption, measured by water present in the filter at the time of the visit, was observed in 78.7 % of households visited by CHWs, 81.1 % of households during the first verification round and then a decrease of nearly 5 % (76.5 %) in the second verification round.

Additional questions were asked of verification survey households only. During both rounds, over 80 % of households reported filling the filter today (44.8 % – 1st VS, 41.8 % – 2nd VS) or yesterday (42.8 % -1st VS, 44.6 % – 2nd VS) with the remainder (12.4 % – 1st VS, 13.6 % – 2nd VS) reporting filtering more than two days ago or not knowing the last time the filter was filled. Additionally households were asked to demonstrate use of the filter. Enumerators recorded performance in meeting up to seven actions. Most households in both rounds (97.5 % – 1st VS, 97.3 % – 2nd VS) were given a rating of sufficient or higher, with nearly 50 % (48.9 %, 43.8 %) receiving excellent ratings. Only 25 households in the 1st round and 5 households in the second round (2.5 % – 1st VS, 2.7 % – 2nd VS) were given a rating of insufficient and thus unable to demonstrate proper usage of the filter.

Households who did not report treating their water during either verification survey round (56 households total), reported this was due to habit (26.9 %), their filter being damaged (16.4 %) and no availability of water in the home (13.4 %). While the 6 verification households who reported using a different treatment method, did so because their filter wasn't working (36.4 %) and they didn't know how to use the filter (27.3 %).

Extensive piloting was conducted to determine the likely least subjective method of determining water volume treated. Quantity of water treated was calculated by the size of the vessel reported used to fill the filter multiplied by the reported number of times the filter was filled each day. This was divided by the number of persons (adults and children) living in the household to yield the liters per person per day (LPPD). Average filtered water volume across the sample, including non-users (0 liter per day) was 1.48 (SD=.80) liters per person per day during the first round and 1.44 (SD=.72) liters per person per day during the second round. The majority of households (81.9 % – 1st VS, 84.2 % – 2nd VS) use filtered water only for consumption with the remaining households (18.1 % – 1st VS, 15.8 % – 2nd VS) using filtered water for additional purposes including cleaning the filter (40.9 % – all VS), washing dishes (29.6 % – all VS) and cooking (18.7 % – all VS). Households reported a 140 % increase in the first round (SD: 139 %) and a 161 % increase in consumption of water from before receiving the filter to after.

Drinking untreated water was reported in 369 of verification survey responses with 33 (2.9 %) households reporting drinking some untreated water at home and 336 (29.3 %) households reporting drinking some untreated water away from home. When drinking water outside of the home, households were primarily traveling (35.4 %), at school (29.2 %) or at work (27.8 %).

While the filter itself has approximately 5.5 liters of storage capacity, 67.8 % of households across both verification rounds report storing additional filtered water. The majority (82.8 %) store in a covered container which is usually a jerry can of various sizes. Households who store water report cleaning their storage container at least once a week (96.8 %) mostly with filtered water (44.9 %) and untreated water (24.2 %). Additionally the safe storage symbol which was promoted through the program to be affixed to any storage containers designated for safe water storage was observed on 89.3 % of containers identified as water storage containers by households.

The primary maintenance task required for the filter is backwashing of the filter membrane. Most verification households (95.5 %) reported backwashing their filter every time they filtered water as advised during household education.

Additional findings include that many (70.9 %) households in the verification sample share water with people outside their household. Of the households that shared water, only 19.7 % reported usually sharing, while the remaining 80.3 % reported sharing sometimes or rarely. Verification households generally did not have negative feedback on how to improve the filter, with most households (69.4 %) reporting no changes to the filter. Other responses included increasing the volume (8.9 %), adding a stand to the bottom of the filter (5.8 %) and providing a cleaning accessory for easier maintenance (4.9 %). Additionally households primarily reported that they liked the filter because it provided clean water (43.7 %), they like the taste of the water (14.2 %), it provides safe water storage (10.6 %) and it saves fuelwood from not having to boil water (10.3 %).

## 8.5   Improved Cookstove Adoption Indicators

92.8 % of households in both the CHW survey (91,704 of 98,804) and the first veri-fication survey (928 of 1,000) reported the EcoZoom stove as their primary cook-stove, with a small decrease to 89.3 % during the second verification round. The next most frequent response was the traditional three stone fire with less than 5 % for the CHW survey and the first verification round (4.9 %) with an increase to 9.6 % during the second verification round. When asked which stove was cooked on during the last cooking event, EcoZoom use reduced to around 80 % of responses (79.2 % CHW, 82.0 % 1st VS, 80.5 % 2nd VS) while the traditional three stone fire increased by less than 15 % for all rounds (14.6 %). Observed EcoZoom use was also lower based on stoves that CHWs and enumerators witnessed cooking on at the time of the household visit (75.2 % CHW, 77.9 % 1st VS, 83.3 % 2nd VS). Additionally, households reported use of the pot skirt, in about 7 out of 10 cooking events during both verification rounds (68.9 % 1st VS, 67.1 % 2nd VS).

The 10 households (0.8 %) between both verification survey rounds which reported not using the EcoZoom stove, reported they didn't know how to use it (23.1 %), it didn't warm the house (23.1 %) or it was difficult to use (15.4 %) as the reported reasons for non-use.

Enumerators performing the verification survey asked households to demon-strate proper cookstove use with each household receiving an internally recorded rating based on number of successful use and maintenance steps completed. Almost all households (98.3 %) received a rating of sufficient to use the EcoZoom stove or better with 79.0 % of households receiving an excellent rating. Only 1.7 % of house-holds received a rating of insufficient for use of the cookstove.

While households reported use of a primary stove, about half the households (48.6 % CHW, 47.5 % 1st VS, 51.3 % 2nd VS) reported usage of other stoves as well. The traditional three stone fire (54.7 % all VS) was the most common supple-mentary stove followed by the *Rondereza* (21.6 % all VS). Based on the number of cooking events reported by each verification household, the EcoZoom was used on average in 86.4 % (SD: 18.4 %) of a household's cooking events during the first verification round and then increased to 92.5 % during the 2nd verification round. The most frequently reported reasons for using another stove included difficulty in finding dry fuelwood to use in the EcoZoom stove (32.2 %), the need to use multiple stoves at one time (24.2 %) and the need to warm the home (15.1 %).

Wood was the primary reported cooking fuel in about 97 % of households for all surveys (97.0 % CHW survey, 97.0 % 1st VS, 96.8 % 2nd VS), though only 90.0 % of households were using wood in observed cooking events by verification survey enumerators. Most verification households reported only collecting wood (74.1 %) while 10.2 % reported both collecting and purchasing wood, and the remainder (15.7 %) only purchasing wood. 92.8 % of households reported storing wood, a highly emphasized part of the education program to promote drying of wet fuel-wood, with most households storing wood inside the home (59.7 %) and a third storing in a separate kitchen (34.4 %).

The majority of households reported cooking outdoors (66.0 % CHW, 76.4 % 1st VS, 75.4 % 2nd VS) with cooking in a doorway (21.5 % CHW survey, 11.5 % 1st VS, 5.9 % 2nd VS) as the next most frequent cooking location. Slightly lower outdoor cooking (62.7 % 1st VS, 61.3 % 2nd VS) was observed when households were cooking at the time of the verification household visits with over a quarter (28.1 % 1st VS, 32.3 % 2nd VS) of households cooking indoors or in a separate kitchen. Households reported cooking indoors fewer times per week than before receiving the EcoZoom stove (7.33 1st VS, 7.23 2nd VS). Primarily households reported cooking indoors because they were getting away from rain (33.8 %) followed by cooking on a stove that could not be moved outdoors (18.7 %), the need to warm the house (12.3 %), security (9.7 %) and habit (9.6 %).

When asked what could be improved on the stove, the majority of verification household's responses were no improvements (60.4 %) with other frequent responses including increasing the size of the stick support (11.9 %), increasing the size of the stove top (7.8 %) and providing a stove that can use multiple fuels (7.1 %). Households additionally reported liking the stove because it cooks fast (32.9 %), reduces fuelwood (30.5 %) and produces less smoke (19.9 %).

## 8.6   Quality Assurance Evaluation

To reinforce the value of household education and interaction, several quality assurance activities were instituted. Before CHWs were allowed to perform household visits alone, a group household visit was conducted with the supervisor to offer feedback and provide clarification for a high quality household visit. CHWs were continually tracked against several metrics including number of surveys per day, average time spent in households and a qualitative evaluation performed by their supervisor. Of 864 CHWs, 774 (89.6 %) evaluations were submitted by DelAgua staff. About a tenth (10.9 %) of CHWs received an excellent rating, three quarters (74.5 %) received a satisfactory rating and the remainder (14.6 %) received an unsatisfactory rating. CHW performance during household visits was evaluated by number of surveys, average survey time and an additional qualitative evaluation performed by staff during one of the CHWs first visits. On average CHWs performed seven household surveys per day, spending 31 min in a household. CHW evaluations improved slightly from the refresher training with under a tenth (9.1 %) of CHWs performing to an unsatisfactory rating, just over 80 % (80.9 %) receiving a satisfactory rating and 10.0 % receiving an excellent rating. Some CHWs receiving unsatisfactory ratings were dismissed.

CHW metrics were again tracked during the follow up household visits, including supervisor evaluations of CHW education performance through visiting households previously visited by CHWs. Supervisors evaluated a CHWs completion of all education tasks including the presence of the hung poster, the sticker placed on an appropriate safe storage container, and bracelets given to households for adopting the products. Additionally households were asked several questions related to retention of key messages and asked to demonstrate use. A score was calculated based on these metrics and CHWs were ranked as excellent (71.6 %), satisfactory (28.1 %) or unsatisfactory (0.3 %) performers. High performing CHWs were given a bonus, satisfactory CHWs were given no bonus, and unsatisfactory performers were reviewed further for dismissal from the program. Evaluated households were selected by the supervisors with CHW's having no prior knowledge as to which specific household might be selected. On average CHWs performed five household visits per day, slightly lower than the initial household survey of seven per day due to the longer time spent in households (46 min).

## 8.7  Discussion

High levels of initial adoption of the water filter and cookstove in the months to a year following distribution of the products in the large scale *Tubeho Neza* program were seen in this program. Similar rates of reported adoption of both the water filter and cookstove (around 90 %) were seen in the Phase One effort implemented two years prior to the large-scale program.

Filtered water quantity increased from the pilot study of 1.27 liters per person per day to 1.63 liters per person per day. The increase may be attributable to increased emphasis in the behavior change program including added messaging about the importance of hydration and specific activities on the household poster which outline how much water should be treated each day in order for the whole family to drink two liters per person day.

Another significant change in the behavior change program was the addition of safe storage messaging. Anecdotal evidence during Phase One suggested that households desired additional storage inside the home and especially while away from the home. In the *Tubeho Neza* program, the majority of households reported storing filtered water with over 80 % storing in a container with a lid, thus emphasizing the importance of the added messaging. Still, about a third of surveyed households reported drinking untreated water while away from the home, mostly while traveling. Given evidence that drinking untreated water, even occasionally, can reduce health benefits of water quality interventions (Brown and Clasen 2012), continued emphasis on the importance of safe storage and exclusive consumption of filtered drinking water should be promoted within the program.

While current repairs and replacements of water filters have been less than 2 % of the total households, long term adoption will likely only be realized if filters are continually maintained in a timely manner with an efficient supply chain. Currently repairs are mostly performed by program staff but in order to create a sustainable maintenance structure, local repairs will be needed. The program is currently training CHWs to perform more repairs and solve maintenance issues before program staff has to travel to individual households or villages to fix issues. Additionally, one of the more frequent repairs is simply from filters being clogged, likely from these households not backwashing the filter enough. More stress will need to be placed on this maintenance task in future trainings to prevent further clogging issues.

While overall reported stove adoption was comparable to the Phase One effort, improvements were made in stove stacking behavior. Reported use of other stoves reduced by over 20 % to about half of households reporting still using other stoves, with percentage of cooking events on the EcoZoom stove in the household increasing by at least 15 %. While these results are promising in moving towards exclusive adoption of improved stoves, they will not be sufficient in meeting the World Health Organization's guidelines for indoor air pollution (WHO 2014) which would involve switching to much cleaner fuels and stoves in order to meet recommendations. However, recent evidence suggests that stove interventions may be evaluated based on both the fuel/stove combination and program usage rates as health gains can be made with lower performing stoves when usage rates are high (Johnson and Chiang 2015).

One suggested solution to address stove stacking is to provide larger households a second improved stove as many households report desiring a second improved cookstove. However, there was no apparent correlation between household size and stove stacking behavior.

Another frequently reported behavior change barrier during the pilot was the inability to cook on the EcoZoom stove when fuel was wet. Wood storage messaging to promote drying of wood before households needed fuel for cooking was added and promoted highly through the education and training materials, resulting in over 90 % of households reporting storing wood and over 65 % having dry wood present in their household at the time of the visit. However the primary reported reason for not only using the EcoZoom stove was still a household's inability to find dry fuel for the EcoZoom.

Rates of outdoor cooking additionally improved from the pilot with 20 % higher frequency of outdoor cooking observed during household visits. A common issue during the pilot was the inability to cook outdoors while it was raining and thus cooking in the doorway as an alternative cooking location was highly emphasized during household visits, where many households reported the doorway as their primary cooking location. The behavior change of cooking outdoors may provide additional important health benefits. The potential for reductions in exposure from cooking outdoors were highlighted in the Phase One independent randomized controlled trial (described below) wherein mean $PM_{2.5}$ concentrations were reduced by 39 % for those cooking indoors on the EcoZoom with further reductions of 73 % when cooking outdoors on the EcoZoom.

Free distribution of health products is often debated, centered around claims that free products do not result in adoption rates needed to realize health benefits. The Tubeho Neza program suggests that free distribution is not a determinant of low adoption. It is plausible that continued engagement in households, enabled by Ministry of Health support and carbon financed revenue, contributed to high adoption rates. Overall, the *Tubeho Neza* program was able to demonstrate high levels of initial adoption and usage of household level water filtration and improved cookstoves at the large scale.

## 8.8   Monitoring Performance

Successful health interventions require extensive monitoring and follow up. Likewise, the carbon crediting model (described below) requires continual evidence of the program function. Within this program, several methods are used to monitor and credit the effort.

Mobile data collection is used for all phases of the program. The data obtained during each campaign is kept in the central database where each record is screened for validity. Valid record submissions are then used for analysis and tracking purposes. Through linking each household and record to the technology identification barcode and GPS coordinates, routines are run to check that filters and stoves are in the households that they have been assigned to. If a filter or stove is found outside of the project area or not in the household that it was originally assigned to the record is flagged and an investigation into the cause of the movement of the technology can be initiated. The centralized database also allows for geospatial analysis of household level data such as performance and usage, which provides feedback for subsequent education and follow-up campaigns.

Through the implementation of mobile data collection DelAgua is able to effectively administer and manage large-scale development programs with expedited technology rollouts, detailed household level education campaigns, and comprehensive technology product tracking.

However, the survey tools are insufficient as exclusive measures of impact. In the case of household level interventions, such as sanitary latrines, household water filters, and improved cookstoves, the measurement methods have relied heavily on surveys and observations. These surveys and other common methods for assessing behavioral practices are known to have certain methodological shortcomings. Surveys often overestimate adoption rates due to reporting bias where the participant is trying to please the surveyor, or recall bias where the participant does not remember the information correctly. This phenomena has been demonstrated in a number of applications, including determinations of poor correlation between observations and self-reported recall of water storage, hand washing and defecation practices (Stanton et al. 1987; Manun'Ebo et al. 1997).

Survey results can also be impacted by errors of interpretation on the part of the informant or the enumerator. Missing data due to participant absences or loss to follow up is another source of systematic bias. Additionally, it is known that the act of surveying can itself impact later behavior (Zwane et al. 2011). Structured observation, an alternative to relying on reported behavior in response to surveys, has also been shown to cause reactivity in the target population (Clasen et al. 2012a). Finally, the subjectivity of the outcome studied can highly influence reporting bias (Wood et al. 2008).

In some instances, it is possible to rely on objective indicators of target behaviors. Programs that implement household water treatment with chlorine, for example, can assess use by testing stored drinking water for chlorine residual. While this is a more objective measure, it yields only infrequent data points and provides information only on water treatment, not actual use of treated water. Moreover, reporting bias can still exist with users chlorinating when they find out a surveyor is visiting. Other options, such as laboratory assays of drinking water quality or hand contamination are costly and unreliable measures of uptake of target behaviors. Additionally, many energy programs use exhaustive lab testing methods to assess specific improved cooking stoves but with little correlation to typical cooking conditions (Roden et al. 2009). An objective, affordable, timely and continuous monitoring system has yet to be realized within the development sector that can meet all of these challenges.

In recent years, there exists an emerging interest in the application of electronic instrumentation to monitor international development programs in an effort to instigate transparency, accountability, and enhance monitoring and evaluation (Ruiz-Mercado et al. 2012; Thomson et al. 2012; Clasen et al. 2012b; Thomas et al. 2013). These monitoring applications span a range of integration strategies such as cell phone based surveys, low-range communication instrumentation, and GSM enabled sensors. Remote sensors are one that engineers have sought to utilize due the rapidly growing broadband network expansion sweeping across the developing world. This is relevant on the African continent where the more than 130 million people lacking water have mobile network coverage, 64 % of whom live in a rural setting where handpumps are widely selected as the appropriate technology (2013).

Within the previously described Phase One effort we collected data from intervention households on product compliance using (i) monthly surveys and direct observations by community health workers and environmental health officers, and (ii) sensor-equipped filters and cookstoves deployed for about two weeks in each household. The adoption rate interpreted by the sensors varied from the household reporting. 90.5 % of households reported primarily using the intervention stove, while the sensors interpreted 73.2 % use. 96.5 % of households reported using the intervention filter regularly, while the sensors interpreted no more than 90.2 %. The sensor-collected data estimated use to be lower than conventionally-collected data both for water filters (approximately 36 % less water volume per day) and cookstoves (approximately 40 % fewer uses per week) (Thomas et al. 2013).

The use of sensors on water filters allowed us to examine objectively the quantity of water being treated in the home and the consistency of filter use. The intrahousehold evaluation of consistent use suggested that most households are not treating enough water to meet their daily needs, potentially indicating non-exclusive use of the filter. This has potential implications for health and environmental impacts and may be considered in further behavior change messaging and impact evaluations. Likewise, any inconsistent use of the cookstoves may be indicative of stove-stacking behavior. As behavior change efforts are modified and expanded, the instrumentation may be an effective tool for evaluating the effectiveness of this messaging.

These results provide additional evidence that surveys and direct observation may exaggerate compliance with household-based environmental interventions. However, the question remains whether the use of these devices influence the behav-

**Fig. 8.4** Cellular enabled sensor installed in a sample of water filters and cookstoves monitor use

ior that is being monitored. Building on this study, we conducted a cluster random-ized trial to assess the behavioral impact of instrumented monitoring on household use of water filters and cookstoves among two groups: (1) Households blinded to (hidden) sensors installed on their water filters and cookstoves (Blinded Arm); and (2) Households informed about visible sensors mounted on water filters and cook-stoves (Open Arm) (Fig. 8.4). A 63 % increase in number of uses of the water filter per week between the groups was observed in week 1, an average of 4.4 times in the open group and 2.83 times in the blind group, declining in week 4 to an insignificant 55 % difference of 2.82 uses in the open, and 1.93 in the blind. There were no sig-nificant differences in the number of stove uses per week between the two groups. For both filters and stoves, use decreased in both groups over four-week installation periods. This study suggests behavioral monitoring should attempt to account for reactivity to awareness of electronic monitors that persists for weeks or more.

The insights gained from these instrumented monitoring studies are relevant only if leveraged toward improved service delivery in the broader context of envi-ronmental health. In this way, the instrument-derived data insights are contextual-ized in health behaviors. In environmental health interventions addressing sanitation, water and indoor air quality, multiple determinants contribute to service utilization. These may include technology selection, technology distribution and education methods, community engagement with behavior change, and duration and magnitude of implementer engagement. Necessarily, funding incentives and performance models must be considered as system design challenges in order to effect service delivery.

## 8.9   Monitoring Health Impacts

The program described in this chapter was accompanied by several health impact randomized controlled trials, funded by the implementer. This section describes the trials, and results available as of 2015 (Fig. 8.5).

## KEY FINDINGS

Preliminary results suggest that after 2 years,
Phase 1 of the programme achieved:

| 46% | 73% | 27.7% |
|---|---|---|
| reduction in diarrhoea odds in children under 5 | reduction in household air pollution for families cooking outdoors | reduction in cookstove emission exposure among children |

Phase 2 of the programme has reached nearly 460,000 people,
including 140,000 children under 5. If these results can be sustained
throughout this phase:

The programme may be expected to save more than

**30**

children's lives a year

&

avert over

**2,500**

disability-adjusted life-years (DALYs) annually

**Fig. 8.5**  Initial Health Impact Results of Tubeho Neza Phase One Program. Morbidity and mortality figures are indicative estimates

## 8.9.1   Phase One

The Phase One program included a five-month cluster randomized trial among 566 households in three pilot villages to assess coverage and use, the impact of the water filter on fecal indicator bacteria in household drinking water and the impact of the stove on fine particulate mater (PM2.5) and carbon monoxide (CO) in reported cooking areas (Rosa et al. 2014). While reported filter use was high (89.2 %), 25 % reported drinking from other sources at least once during five follow-up visits; filter-mounted sensors also showed self-reports to exaggerate use (Thomas et al. 2013). Overall, the intervention was associated with a 97.5 % reduction in mean faecal indicator bacteria (0.5 vs. 20.2 TTC/100 mL, p<0.001). Two-thirds (66.7 %) of intervention households identified the intervention stove as their main cooking stove, but only 23.3 % of intervention households reported that their main cooking area was outdoors. Overall, the stoves were associated with a 48 % reduction of 24-h PM2.5 concentrations in the cooking area (0.485 mg/m$^3$ and 0.267 mg/m$^3$, p=0.005). The reduction was 37 % for those cooking indoors (p=0.08) and 73 % for those cooking outdoors (p<0.001) (Rosa et al. 2014).

Following the Phase One RCT, nine of the non-RCT pilot villages were matched with control villages and followed for an additional 12 months to assess longer-term intervention uptake and to test methods for assessing exposure and health outcomes for a larger scale health impact evaluation. Households were surveyed once in round 1 12–18 months after first receiving the intervention, and a second time 6 months later. In round 1, 113 intervention and 156 control households were enrolled and surveyed; 91 of these intervention and 144 of these control households were followed up approximately 6 months later in round 2.

In both rounds, 82 % of intervention households still had the filter that was reported to be working properly, and more than 95 % of intervention houses still had the EcoZoom stove. In both rounds, 90 % of intervention households reported currently using the filter, 95 % of whom within the previous 2 days; usage was similar with the EcoZoom stove, with at least 87 % of intervention households reporting current usage in both rounds, 95 % of whom within the last 2 days. Among intervention households that had drinking water in the house at time of visit, 74 % in round 1 reported the water had been treated (99 % using the filter), and 78 % in round 2 (97 % with the filter). Using combined data from both rounds, the filters were associated with a 79 % reduction in mean faecal indicator bacteria (geometric mean 1.3 vs. 6.3 TTC/100 mL, p<0.001. The odds of having contaminated drinking water (>=1 TTC/100 mL) were 3.7 times higher in the control arm than the intervention arm (p<0.001). In order to assess the impact of the EcoZoom stove on reducing exposure to harmful cooking smoke, we measured personal exposure to particulate matter (PM2.5) in cooks and children under 5 for a 48-h period. Intervention cooks had a reduced exposure of 26.5 % (p=0.01) compared to controls (predicted mean 198.9 vs. 270.6 µg/m$^3$), while intervention children under 5 had a reduced exposure of 27.7 % (p<0.001, predicted mean 219.1 vs. 303.0 µg/m$^3$). While this study was not designed to assess health impact, primary caretakers were asked to report on their child's health in the previous 7 days (n = 338 children in round 1, 305 in round 2). Odds of diarrhoea in intervention children were 46% less than control children (OR 0.54, p=0.04). No impact has been observed in the prevalence of proxy indicators for ALRI. This study found high uptake and sustained use of a household water filter and advanced cookstove 1–2 years following intervention delivery, with evidence of water quality improvement, smoke exposure reductions, and improvements in child health outcomes.

### 8.9.2   Phase Two

Within the Phase Two program, a sector-level cluster-randomized controlled trial was conducted to assess the impact of the intervention on health outcomes using records maintained by the clinical and CHWs (the "clinic-level RCT"). At the same time, we randomly selected 87 villages from each arm of the sector-level RCT for nested village-level RCT where we could assess coverage, uptake (use), exposure and other measures of health outcomes (reported, CHW recorded, instrumented and potential blood-based biomarkers) (the "village-level RCT"). Following our baseline study, the implementer delivered the intervention to approximately 100,000 eligible households within the 72 intervention sectors (September – December 2014).

### 8.9.3   Clinic-Level RCT

The main objective of the clinic-level RCT is to assess the impact of the intervention on health. The study is among the largest RCTs ever conducted in environmental health and includes all 96 sectors (~3700 villages and approximately 140,000 Ubudehe 1 and 2 households) in Western Province. The main advantages of this trial over previous research are (i) its use of more objective health data drawn from clinical records with no potential bias arising from self-reported conditions reported on multiple visits by enumerators; (ii) its ability to investigate a wide variety of health outcomes potentially related to water quality and HAP, including not only ALRI and diarrhea (cause mortality, mortality attributed to respiratory disease or diarrheal disease, tuberculosis, hypertension, low birthweight, premature birth and stillbirth, and (iii) its unprecedented size, which offers the potential to assess the impact of the intervention on mortality. We worked closely with the MOH on data entry using standard forms; they are enthusiastic about our using routinely collected health data for research. The study is powered to detect a 10 % difference in primary outcomes between intervention and control groups after adjusting for clustering. No baseline was necessary, since the Field teams will be visiting clinics to extract the relevant data (February-May 2016). After an analysis period (June-September 2016), we will write up and submit the results of the clinic-level trial by the end of 2016.

### 8.9.4   Village-Level RCT

The village-level RCT will also provide data on health outcomes, including diarrhea and pneumonia. The main objective of the village-level RCT is to assess the impact of the intervention and HAP and faecal contamination of drinking water—the main exposures that the intervention aims to mitigate. A sub-study is also designed to investigate possible biomarkers enteric and respiratory disease in an effort to develop more objective criteria for assessing these health disorders and the interventions designed to prevent them.

For the village-level RCT, we enrolled 1582 households with children <5 from the 174 study villages, evenly distributed between intervention and control arms. At baseline, we collected extensive information from study participants on demographics, water sources and management practices, cooking fuels and cooking practices. Diarrhea is assessed based on 7-day self-reports; ALRI is assessed using World Health Organization (WHO) and Integrated Management of Child Illness (IMCI) criteria for pneumonia case identification in resource-limited settings. This includes severity indices that incorporate cough, difficulty breathing and rapid respiration (Puumalainen et al. 2008). The village-level RCT is powered to detect a 25 % difference in diarrhea or pneumonia.

A sub-sample of two households in each study village was randomly selected to undergo extended health and exposure evaluation (sub-study households). This includes an extensive panel of physiologic measurements to assess blood pressure, carboxyhemoglobin (COHb) concentrations (through pulse oximetry and exhaled CO), $O_2$ saturation ($SpO_2$) and levels of various biomarkers of HAP exposure, enteric infection and ALRI. Blood pressure is assessed among main cooks in these households using a blood pressure monitor with cuff. Eligible adults for this sub-study consist of women at least 16 years of age who are identified as the main cook for the household and children under-5 years-old who live in selected households. Personal level gravimetric PM2.5 exposures are obtained cumulatively over 48 h from the main cook and a child between 1.5 and 4 years using a wearable pump/filter that includes a light sensor to assess time-specific location and compliance. This also allows us to explore dose-response relationships and to contribute to the limited knowledge on the relationship between exposure and disease.

Dried blood spot samples were obtained at baseline and the second follow up round in order to assess the utility of various biomarkers and as indicators of systemic inflammation, personal exposure to PM2.5, seroconversion to enteric pathogens and specific mechanistic disease processes. Biomarkers offer the potential for improving the consistency of diagnoses of these diseases and the exposures from contaminated air and drinking water that contribute to their high prevalence in low-income countries; in addition, they can help inform and improve the reliability of standardized disease outcome classifications. We anticipate assessing levels of various inflammatory biomarkers (including interleukin (IL)-6, IL-8, IL-10, tumor necrosis factor-alpha (TNF-α) and C-reactive protein (CRP) for the cookstove component; antibodies against antigens for *Giardia spp.*, *Cryptosporidium spp.*, *E. coli* and other enteric pathogens for our seroconverson study; and biomarkers reflecting intestinal dysfunction and permeability to assess environmental enteropathy.

## 8.10   Incentivizing Impacts

In the context of this program, the for-profit implementer, DelAgua, is contributing directly to national health policy goals, and is able to finance the effort through the carbon credits generated and sold. These carbon credits are tied nearly exclusively and proportionally to demonstrated product adoption and correct use – the carbon credits are calculated based on on-going monitoring of stove and filter use (Kremer et al. 2008). The company is commercially aligned with prioritizing adoption, and can choose to invest in continuous engagement. In this program, DelAgua secured external, third party investors, motivated by both the development impacts and the projected financial returns.

This section provides an overview of how carbon credit markets are applied to these kinds of environmental health program as a profitable business venture, and extends the concept to crediting and monetizing health impacts directly without the carbon finance intermediary.

Carbon finance markets facilitate the reduction of greenhouse gas emissions worldwide through economic incentives, while allowing cleaner economic development to take place. Each emission reduction credit represents the non-emission of one tonne of carbon dioxide into the atmosphere. The carbon credits generated under the CDM help Kyoto Protocol Annex I countries to meet their binding targets, and can be traded in the marketplace. However, the carbon markets have yet to be well utilized to finance the distribution of humanitarian technologies in the least developed countries, particularly in Africa. Although the CDM is a multi-billion dollar industry, fewer than two percent of projects are registered in African nations (Williams and Murthy 2013) (Fig. 8.6).

Depending on the project location, structure, methodology and registration mechanism employed, a water treatment and/or cookstove program can earn between approximately 1/2 and 5 carbon credits per household, per year. The carbon credits earned are a function of the approved methodology, referenced to a baseline condition and the current performance of the program, as audited by independent firms. The reported reductions are then issued by the registration authority and are then sold to buyers. Because the carbon credits are issued in proportion to the present adoption and proper use of intervention technologies, this encourages sustained engagement by the program implementer and creates a pay-for-performance model.

Carbon credits associated with cookstoves are tied to reduced wood fuel use. In the case of water filters, the credits rely on a concept called 'suppressed demand', created by the carbon crediting authorities, that presumes a wood fuel use 'demand' associated with treating water by boiling. In reality, only a minority of households do boil their water, while most drink untreated water. This methodological construct has been extensively debated, and is reviewed in a following chapter.

Separate from the technical premise of suppressed demand for water treatment, it is important to recognize the political considerations that led to the suppressed demand approach. With less than 2 % of CDM programs active in Africa, there was a clear skew in favor of developing economies rich enough to realize the tremendous

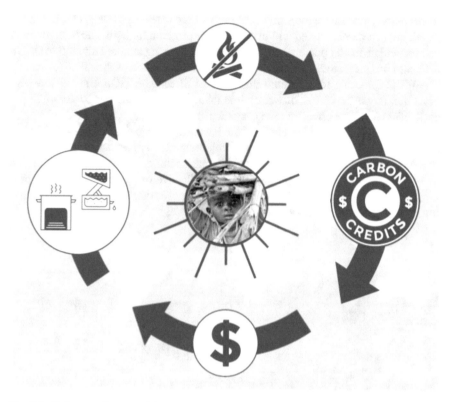

**Fig. 8.6** Carbon credits earned from household water filters and cookstoves are associated with continued demonstration of adoption, thereby creating a closed loop funding incentive to maintain the program

demand for energy, while poorer countries, with the same demand for the health and welfare benefits from increased energy usage, were left out of the market.

On the cost side, household interventions like cookstoves and water filters can be procured for roughly $25 each. In Rwanda, the logistics of getting a device into a household may cost another $5. Annually, servicing costs, inclusive of household visits and behavior change messaging, plus the amortized replacement of the device, may be roughly $10.

Assuming a 85 % sustained adoption rate (higher than many programs, but feasible in a well supervised intervention), a water filter may generate, under the CDM, roughly one credit per year, and a cookstove two credits per year. A typical open market price for UN carbon credits in recent years has been greater than $5, while some projects can fetch double or triple this price based on the nature and credibility of the intervention. Carbon credit buyers may be banks, energy companies, brokers, or sovereign nations who require credits for either regulatory compliance or voluntary social responsibility efforts, or both.

In addition to the environmental benefits represented by the carbon credits, there is an emerging alignment between monitored health impacts, calculations of units of heath impact (Averted Disability Adjusted Life Years – ADALYs), and, finally, payments associated with demonstrated ADALYs (See Chapters 10 and 11). Using the health impacts demonstrated within the Rwanda program for both diarrhea reduction and particulate matter personal exposure reduction among children under 5, ADALY estimates are presented. The value of these ADALYs is tied to the GDP per capita. The WHO Choosing Interventions that are Cost Effective (CHOICE) guideline suggests that any intervention that costs less than three times the per capita GPD per each ADALY is cost effective (See Chapter 10).

In Fig. 8.7, a simplified analysis of the Rwanda program is presented. While the cost, carbon credit, health and benefit calculations are complex and more conservative or optimistic adjustments to any single parameter can change the bottom line margin, the numbers presented are roughly reflective of the nature of these programs. This table is presented for illustrative purposes only, as each parameter has intrinsic and extrinsic complexities (Fig. 8.7).

| | One Water Filter | One Cookstove | 100,000 Households |
|---|---|---|---|
| Installation Cost | $30 | $30 | $6,000,000 |
| Annual Cost | $10 | $10 | $2,000,000 |
| 10-Year Total Cost | $130 | $130 | $26,000,000 |
| Annual Carbon Credits | 1 | 2 | 300,000 |
| Annual Carbon Credit Revenue ($5 sales price) | $5 | $10 | $1,500,000 |
| Annual Averted Morbidity and Mortality (ADALY) | 0.02 | 0.01 | 3,000 |
| 10-Year ADALY | 0.2 | 0.1 | 30,000 |
| 10-Year Cost / ADALY | $650 | $1,300 | $578 |
| Annual Health Benefit Value 1 ADALY Value = 3 x GDP / capita ($640 Rwanda, 2013) | $38.4 | $19.2 | $5,760,000 |
| 10-Year Carbon Value | $50 | $100 | $15,000,000 |
| 10-Year Health Benefit Value | $384 | $192 | $57,600,000 |
| Total Net Value | $496 | $258 | $46,600,000 |

**Fig. 8.7** Indicative estimates of costs and revenue associated with a carbon credit generating water filter and cookstove intervention

## 8.11    Implications

In this program, high levels of uptake and continued use of water filters and improved cookstoves were found. This outcome may be explained through the benefits of continued engagement with communities, enabled by the nature of the program funding.

The extensive logistical and behavior change messaging components of this program require sustained funding. Donation based non-profits are not providing services to the target populations serviced by this program. The operational feedback applied is designed to recuperate invested costs by the generation and sale of carbon credits associated with the proportion of the intervention that continues to demonstrate successful behavior change. The outcomes observed to-date support the business model in that high adoption rates have correlate to carbon credit generation and sale sufficient to generate sustainable revenue that presently allow continued household environmental health service delivery.

## References

Aboud FE, Singla DR (2012) Challenges to changing health behaviours in developing countries: a critical overview. Soc Sci Med 75(4):589–594. doi:10.1016/j.socscimed.2012.04.009

ARC (2012) Results of water boiling test 41 testing of the EcoZoom Dura stove. Aprovecho Research Center, Cottage Grove

Barstow CK, Ngabo F, Rosa G, Majorin F, Boisson S, Clasen T, Thomas EA (2014) Designing and piloting a program to provide water filters and improved cookstoves in Rwanda. PLoS One 9(3):e92403. doi:10.1371/journal.pone.0092403

Becker M (1974) The health belief model and personal health behavior. Health Educ Monogr 2:324–473

Black RE, Morris SS, Bryce J (2003) Where and why are 10 million children dying every year? Lancet 361(9376):2226–2234. doi:10.1016/S0140-6736(03)13779-8

Brown J, Clasen T (2012) High adherence is necessary to realize health gains from water quality interventions. PLoS One 7(5):e36735. doi:10.1371/journal.pone.0036735

CIA (2012) CIA world factbook. Central Intelligence Agency, Washington, DC

Clasen T, Schmidt WP, Rabie T, Roberts I, Cairncross S (2007) Interventions to improve water quality for preventing diarrhoea: systematic review and meta-analysis. BMJ 334(7597):782. doi:10.1136/bmj.39118.489931.BE

Clasen T, Naranjo J, Frauchiger D, Gerba C (2009) Laboratory assessment of a gravity-fed ultra-filtration water treatment device designed for household use in low-income settings. Am J Trop Med Hyg 80(5):819–823

Clasen T, Boisson S, Routray P, Cumming O, Jenkins M, Ensink JH, Bell M, Freeman MC, Peppin S, Schmidt WP (2012a) The effect of improved rural sanitation on diarrhoea and helminth infection: design of a cluster-randomized trial in Orissa, India. Emerg Themes Epidemiol 9(1):7. doi:10.1186/1742-7622-9-7

Clasen T, Fabini D, Boisson S, Taneja J, Song J, Aichinger E, Bui A, Dadashi S, Schmidt WP, Burt Z, Nelson KL (2012b) Making sanitation count: developing and testing a device for assessing latrine use in low-income settings. Environ Sci Technol 46(6):3295–3303. doi:10.1021/es2036702

EUEI (2009) Biomass Energy Strategy (BEST), Rwanda – volume 3 rural supply and demand. European Union Energy Initiative, Brussels

Ezzati M, Kammen D (2001) Indoor air pollution from biomass combustion and acute respiratory infections in Kenya: an exposure-response study. Lancet 358(9282):619–624

Fewtrell L, Kaufmann RB, Kay D, Enanoria W, Haller L, Colford JM Jr (2005) Water, sanitation, and hygiene interventions to reduce diarrhoea in less developed countries: a systematic review and meta-analysis. Lancet Infect Dis 5(1):42–52. doi:10.1016/S1473-3099(04)01253-8

Figueroa M, Kinaid D (2010) Social, cultural and behavioral correlates of household water treatment and storage. Johns Hopkinds Bloomberg School of Public Health, Baltimore

GSMA (2013) GSMAi number of unique subscribers. GSMA, London

Gundry S, Wright J, Conroy R (2004) A systematic review of the health outcomes related to household water quality in developing countries. J Water Health 2(1):1–13

IMF (2012) World economic outlook 2012 – coping with high debt and sluggish growth. International Monetary Fund, Washington, DC

Johnson MA, Chiang RA (2015) Quantitative guidance for stove usage and performance to achieve health and environmental targets. Environ Health Perspect 123(8):820–826. doi:10.1289/ehp.1408681

Kirby M (2013) Water quality testing in 30 districts of Rwanda (Unpublished), London School of Hygiene and Tropical Medicine (in preparation)

Kremer M, Miguel E, Mullainathan S (2008) Source dispensers and home delivery of chlorine in Kenya. Innovations for poverty action. Poverty Action Lab, Cambridge

Lantagne D, Khush R, Montgomery M (2012) A toolkit for monitoring and evaluating household water treatment and safe storage programmes. World Health Organization, Geneva

Lim Y, Kim JY, Rich M, Stulac S, Niyonzima JB, Smith Fawzi MC, Gahire R, Mukaminega M, Getchell M, Peterson CW, Farmer PE, Binagwaho A (2010) Improving prevention of mother-to-child transmission of HIV care and related services in eastern Rwanda. PLoS Med 7(7):e1000302. doi:10.1371/journal.pmed.1000302

Liu L, Johnson HL, Cousens S, Perin J, Scott S, Lawn JE, Rudan I, Campbell H, Cibulskis R, Li M, Mathers C, Black RE, Child Health Epidemiology Reference Group of WHO, UNICEF (2012) Global, regional, and national causes of child mortality: an updated systematic analysis for 2010 with time trends since 2000. Lancet 379(9832):2151–2161. doi:10.1016/S0140-6736(12)60560-1

Logie DE, Rowson M, Ndagije F (2008) Innovations in Rwanda's health system: looking to the future. Lancet 372(9634):256–261. doi:10.1016/S0140-6736(08)60962-9

Manun'Ebo M, Cousens S, Haggerty P, Kalengaie M, Ashworth A, Kirkwood B (1997) Measuring hygiene practices: a comparison of questionnaires with direct observations in rural Zaire. Trop Med Int Health 2(11):1015–1021

Masera O, Edwards R, Arnez CA, Berrueta V, Johnson M (2007) Impact of patsari improved cookstoves on indoor air pollution in Mi-choacan, Mexico. Energy Sustain Dev 11:45–56

Narajo J, Gerba C (2011) Assessment of the LifeStraw family unit using the World Health Organization guidelines for "evaluating household water treatment options: health-based targets and performance specifications". University of Arizona, Department of Soil, Water and Environmental Science, Tucson

Niringiye A, Ayebale C (2012) Impact evaluation of the Ubudehe programme in Rwanda: an examination of the sustainability of the Ubudehe programme. J Sustain Dev Africa 14(3):141–153

Onda K, LoBuglio J, Bartram J (2012) Global access to safe water: accounting for water quality and the resulting impact on MDG progress. Int J Environ Res Public Health 9(3):880–894. doi:10.3390/ijerph9030880

Peletz R, Simunyama M, Sarenje K, Baisley K, Filteau S, Kelly P, Clasen T (2012) Assessing water filtration and safe storage in households with young children of HIV-positive mothers: a randomized, controlled trial in Zambia. PLoS One 7(10):e46548. doi:10.1371/journal.pone.0046548

Puumalainen T, Quiambao B, Abucejo-Ladesma E, Lupisan S, Heiskanen-Kosma T, Ruutu P, Lucero MG, Nohynek H, Simoes EA, Riley I, ARIVAC Research Consortium (2008) Clinical case review: a method to improve identification of true clinical and radiographic pneumonia in children meeting the World Health Organization definition for pneumonia. BMC Infect Dis 8:95. doi:10.1186/1471-2334-8-95

Roden C, Bond T, Conway S, Pinel AB, MacCarty N (2009) Laboratory and field investigations of particulate and carbon monoxide emissions from traditional and improved cookstoves. Atmos Environ 43:1170–1181

Rogers E (2003) Diffusion of innovations, 5th edn. Free Press, New York

Rosa G (2012) Water quality testing in 11 districts in Rwanda (Unpublished), London School of Hygiene and Tropical Medicine (in preparation)

Rosa G, Majorin F, Boisson S, Barstow C, Johnson M, Kirby M, Ngabo F, Thomas E, Clasen T (2014) Assessing the impact of water filters and improved cook stoves on drinking water quality and household air pollution: a randomised controlled trial in Rwanda. PLoS One 9(3):e91011. doi:10.1371/journal.pone.0091011

Rosenstock IM, Strecher VJ, Becker MH (1988) Social learning theory and the health belief model. Health Educ Q 15(2):175–183

Ruiz-Mercado I, Canuz E, Smith KR (2012) Temperature dataloggers as stove use monitors (SUMs): field methods and signal analysis. Biomass Bioenergy 47:459–468. doi:10.1016/j.biombioe.2012.09.003

Rwanda-MOH (2010) Rwanda community based health insurance policy. Ministry of Health, Kigali

Rwanda Ro (2011) Rwanda demographic and health survey 2010. National Institute of Statistics of Rwanda, Kigali

Rwanda-MOH (2012) Third health sector strategic plan July 2012- June 2018. Ministry of Health, Kigali

Saksena P, Antunes AF, Xu K, Musango L, Carrin G (2011) Mutual health insurance in Rwanda: evidence on access to care and financial risk protection. Health Policy 99(3):203–209. doi:10.1016/j.healthpol.2010.09.009

Smith KR, Samet JM, Romieu I, Bruce N (2000) Indoor air pollution in developing countries and acute lower respiratory infections in children. Thorax 55(6):518–532

Stanton BF, Clemens JD, Aziz KM, Rahman M (1987) Twenty-four-hour recall, knowledge-attitude- practice questionnaires, and direct observations of sanitary practices: a comparative study. Bull World Health Organ 65(2):217–222

Stats U (2012) Rwanda summary statistics. United Nations, New York

Thomas E (2012) Leveraging carbon financing to enable accountable water treatment programs. Global Water Forum. http://www.globalwaterforum.org/2012/09/23/leveraging-carbon-financing-to-enableaccountable-water-treatment-programs/

Thomas E, Barstow C, Rosa G, Majorin F, Clasen T (2013) Use of remotely reporting electronic sensors for assessing use of water filters and cookstoves in Rwanda. Environ Sci Tech 47:13602–13610

Thomson P, Hope J, Foster T (2012) GSM-enabled remote monitoring of rural handpumps: a proof-of-concept study. J Hydroinf 14(4):29–39

UNCDM (2013) DelAgua public health program in Eastern Africa, 2012. United Nations Clean Development Mechanism, New York

UNICEF (2015) Committing to child survival: a promise renewed. United Nations Children's Fund, New York

WHO (1997) Guidelines for drinking water quality. In: Surveillance and control of community supplies, vol 3, 2nd edn. World Health Organization, Geneva

WHO (2010) Indoor air pollution and health. World Health Organization, Geneva

WHO (2011) Evaluating household water treatment options: health based targets and microbiological performance specifications. WHO, Geneva

WHO (2014) World Health Organization. http://www.who.int/healthinfo/global_burden_disease/metrics_daly/en/

Williams M, Murthy S (2013) Reconciling the carbon market and the human right to water: the role of suppressed demand under clean development mechanism and the gold standard. Environ Law 43:517–562

Wood L, Egger M, Gluud LL, Schulz KF, Juni P, Altman DG, Gluud C, Martin RM, Wood AJ, Sterne JA (2008) Empirical evidence of bias in treatment effect estimates in controlled trials with different interventions and outcomes: meta-epidemiological study. BMJ 336(7644): 601–605. doi:10.1136/bmj.39465.451748.AD

Wright J, Gundry S, Conroy R (2004) Household drinking water in developing countries: a systematic review of microbiological contamination between source and point-of-use. Trop Med Int Health 9(1):106–117

Zwane AP, Zinman J, Van Dusen E, Pariente W, Null C, Miguel E, Kremer M, Karlan DS, Hornbeck R, Gine X, Duflo E, Devoto F, Crepon B, Banerjee A (2011) Being surveyed can change later behavior and related parameter estimates. Proc Natl Acad Sci U S A 108(5): 1821–1826. doi:10.1073/pnas.1000776108

# Chapter 9
# A Critical Review of Carbon Credits for Household Water Treatment

**James M. Hodge and Thomas F. Clasen**

**Abstract** Household water treatment (HWT) provides a means for vulnerable populations to take charge of their own drinking water quality as they patiently wait for the pipe to finally reach them. In many low-income countries, however, promoters have not succeeded in scaling up the intervention among the target population or securing its consistent and sustained use. Carbon financing can provide the funding for reaching targeted populations with effective HWT solutions and the incentives to ensure their long-term uptake. Nevertheless, programs have been criticized because they do not actually reduce carbon emissions. We summarize the background and operation of carbon financing of HWT interventions, including the controversial construct of "suppressed demand". We agree that these programs have limited potential to reduce greenhouse gas emissions and that their characterization of trading "carbon for water" is misleading. Nevertheless, we show that the Kyoto Protocol expressly encouraged the use of suppressed demand as a means of allowing low-income countries to benefit from carbon financing provided it is used to advance development priorities such as health. We conclude by recommending changes to existing criteria for eligible HWT programs that will help ensure that they meet the conditions of microbiological effectiveness and actual use that will improve their potential for health gains.

**Keywords** Carbon credits • Water filter • Household water treatment • Suppressed demand

Adapted with permission from "Carbon Financing of Household Water Treatment: Background, Operation and Recommendations to Improve Potential for Health Gains", *Environmental Science and Technology, 48 (12) DOI:* 10.1021/es503155m. 2014. American Chemical Society.

J.M. Hodge, JD, MPH • T.F. Clasen, JD, Ph.D. (✉)
Rollins School of Public Health, Emory University,
1363B Markan Dr NE, Atlanta, GA 30306, USA
e-mail: tclasen@emory.edu

## 9.1 Introduction

There are currently two parallel systems of carbon trading: a compliance market and a voluntary market (Arnoldus and Bymolt 2011). Under the Kyoto Protocol, developed countries are required to set emission reduction targets for several greenhouse gases (GHGs) including $CO_2$. The UN Clean Development Mechanism (CDM) was created to allow countries with emission reduction targets to more efficiently comply by purchasing credits generated by GHG reduction projects in developing countries to offset emissions in excess of their cap. Credits purchased from such projects are added to the cap set under the Kyoto Protocol (Arnoldus and Bymolt 2011). As of December 2013, more than 7500 projects had been registered under the CDM and approximately 1.4 billion credits had been issued (CDM 2013a). The number of projects registered in 2013 dropped significantly compared to 2012 but the UNFCCC has renewed the CDM under the second Kyoto Protocol commitment period and ensured its continuation until at least 2020 (CDM 2013a).

The voluntary market comprises several certification organizations, but only The Gold Standard (GS) currently has a methodology for issuing credits for household water treatment programs. Unlike in the compliance market, in the voluntary market, anyone from individuals to multinational corporations seeking to offset carbon emissions can purchase carbon credits, but the majority of voluntary credits are purchased by corporations wishing to voluntarily offset the carbon they are emitting with corporate social responsibility as the main motivating factor (Peters-Stanley and Yin 2013). In 2012, transactions of credits on the voluntary market exceeded 100 million tons of carbon offsets and though market value has decreased, the average price for voluntary credits is higher than that for CDM credits {Peters-Stanley, 2013 #785}. Since its founding in 2004, the GS has registered approximately 1000 projects and issued 24 million credits (Kouchakji).

Significantly, however, a stated goal of the Kyoto Protocol in addition to its reducing carbon emissions was to assist developing countries "in achieving sustainable development" (UN 1998). In essence, the policy behind the CDM is that it is cheaper and more efficient to reduce emissions in developing countries than in industrialized countries. Therefore, the CDM is designed to encourage development projects in non-Annex I (low-income) countries funded by the carbon credits that are sold to carbon emitters or governments in Annex I countries (UNFCCC 2012a).

The CDM and the GS have specified criteria and detailed procedures for registering projects, verifying emission reductions and certifying credits. Once projects are approved and validated, the project promoter is required to monitor the emission reductions and other requirements and demonstrate the level achieved. After the emissions are verified and approved, certified emission reductions (CERs) or voluntary emission reductions (VERs) are issued to the project promoter who can sell them on established markets. Thus, if the projects can generate enough carbon credits under either standard, they have the potential to be self-funding. Recent instability in the carbon credit markets may have reduced the financial benefits but at this time, projects are still ongoing and credits are still being generated and sold.

### 9.1.1   Carbon Credits for Water Treatment

The CDM and the GS each have approved methodologies for reducing emissions by providing low- or no- GHG emitting water treatment technologies. While the CDM has separate methodologies for large- and small-scale programs, only the small-scale methodology (designated "AMS-III.AV") is currently being used (2013b). The GS has one methodology related to water treatment, Technologies and Practices to Displace Decentralized Thermal Energy Consumption Methodology (2011). Both the CDM and GS methodologies explain how to define and calculate the baseline emissions level and the project emissions level in order to calculate the emission reductions and thus the number of credits generated.

### 9.1.2   CDM Methodology- Low Greenhouse Gas Emitting Safe Drinking Water Production Systems

The CDM small-scale methodology includes interventions to distribute water treatment systems and is applicable to projects that reduce up to 60 kt $CO_2$ equivalent per year. A wide range of point-of-use (POU) or point-of-entry (POE) treatment technologies including water filters, UV disinfection devices, solar disinfection techniques, photocatalytic disinfection, pasteurization, chlorination, or combined treatment approaches (e.g. flocculation plus disinfection) can be used to displace boiling under this methodology (2013b). Significantly, the technologies chosen are only required to meet the WHO "interim" guidelines for assessing the microbiological performance of HWT technologies (WHO 2011), or relevant national standards (2013b).

The CDM methodology is available only in locations where there is no existing public distribution network of safe drinking water; it loses applicability if, during the project period, a distribution network is installed (2013b). Projects using this methodology can be in either rural or urban areas in two defined cases: (i) areas where the proportion of the population using an "improved drinking water source" (as defined by the WHO/UNICEF Joint Monitoring Programme for Water Supply and Sanitation) is equal to or less than 60 % (Case 1); or (ii) areas where the use of an improved drinking water source is greater than 60 % (Case 2) (CDM 2013b). There are currently fewer than 20 countries that meet this condition for Case 1 projects (UNICEF 2013; Williams and Murthy 2013).

The baseline level of emissions under this methodology "assumes that fossil fuel or non-renewable biomass (NRB) is used to boil water as means of water purification in the absence of the project activity." (CDM 2013b). However, for Case 2, the project population is adjusted based on the percentage for which it can be demonstrated that the practice of water purification would have been boiling (CDM 2013b). Furthermore, only water purified for drinking purposes may be included in the baseline calculation and the quantity that may be included is capped at 5.5 l per person per day (CDM 2013b). Water quality is required to be monitored on a sample basis during the monitoring and verification process. Laboratory testing or official notifications are required to demonstrate that the application of the project technology meets the WHO interim performance standard or a relevant national guideline (CDM 2013b).

It is also important to note that this methodology does not factor in life-cycle emissions of the treatment technology. There is a parameter to account for $CO_2$ emissions generated by the operation of the treatment technology, but no requirement that the energy consumed or emissions generated by the production of the treatment technology and transport to the project area be considered despite such embodied energy having the potential to be significant (Held and Zhang 2013).

### 9.1.3 GS Methodology: Technologies and Practices to Displace Decentralized Thermal Energy Consumption

The GS methodology applies to displacement of water boiling as a purification technique with "'zero emission technology' that provides safe water, e.g. gravity household water filters, borehole pumps (not fossil fuel driven) and their repair/maintenance/operation, ultraviolet radiation treatment, chlorine tablets, etc." (GSF 2011). Under the terms of the program, credits will only be given for end users that either boil water or use unsafe water at the time of implementation (GSF 2011). The amount of treated water that can be used in calculating the baseline and emission reductions is 7.5 l per person per day. This cap includes the quantity of safe water used in the project scenario "for all purposes where contaminated water would imply a health or livelihood risk" and thus is more expansive than the cap under the CDM AMS.III-AV methodology which is limited to drinking water (GSF 2011).

Under this methodology, water quality is a required parameter to be monitored throughout the crediting period of the project. Quality testing can be conducted in the field or by transportation to a laboratory and in either case, testing procedures are required to be fully described in monitoring reports; however, testing methods are not prescribed (GSF 2011). In contrast to the CDM methodology, projects approved after January 2014 must employ technologies that either meet the relevant water quality standards of the host country or, if such standards do not exist, the WHO "highly protective" performance standard (GSF). An exception to this requirement permits the use of treatment technologies that only meet the 'interim' standard if the project proponent can demonstrate that the project area has no waterborne pathogens of the class that the treatment technology fails to remove (GSF). Projects under this standard are also explicitly required to consider emissions generated by project activities including upstream emissions from manufacture and transport of project materials if those emissions are expected to be greater than 5 % of the overall emissions (GSF 2011).

### 9.1.4   Suppressed Demand in the CDM and GS

"Suppressed demand" is an attempt to recognize that there is a quantifiable level of energy demand that is not being attained due to a variety of impeding factors including poverty and underdevelopment. It has its basis in the modalities and procedures created to govern the CDM that expressly provide that "the baseline may include a scenario where future anthropogenic emissions by sources are projected to rise above current levels, due to the specific circumstances of the host party." (CDM 2012; UNFCCC 2006). In other words, the CDM contemplated that the baseline level from which emission reductions will be calculated should be grossed up in settings where it is currently suppressed by underdevelopment.

The notion of suppressed demand has been widely criticized, both generally and in the context of using carbon credits to finance the distribution of HWT. The main criticism is that the application of suppressed demand results in "avoided emissions" rather than reduced emissions and has no impact on reducing overall emissions. Because the credits generated by suppressed demand projects are then sold on to emitters in developed countries, it results in polluters purchasing offsets that do not directly correspond to any actual reduction in carbon emissions.

These criticisms are valid. However, the reason for inventing the construct of suppressed demand was not to mitigate future GHG emissions, but to allow LDC's a means of participating in the carbon credit mechanism. Because such countries are relatively low emitters, they are unlikely to be able to host financially viable carbon credit projects. Suppressed demand provides a bootstrap for their participation. As stated in the CDM guidelines, "[i]f suppressed demand were not included, baseline emissions would be so small that project activities would become unattractive under the CDM due to the small number of [certified emission reductions] generated." (UNFCCC 2012b).

This problem is evident in the current distribution of GHG emissions and registered CDM projects. Low-income countries account for less than 5 % of global GHG emissions with per capita emissions of only approximately 0.2 tons. By comparison in 2010, the United States per capita emissions of just carbon were 19.9 tons and China's per capita emissions were 6.2 tons (World Bank). In the absence of suppressed demand, baseline emission levels in Low-income countries are so low that the amount of credits generated through actual reductions will be insufficient to recover the costs of implementation. Even with suppressed demand as a possible mechanism for making such programs viable, Low-income countries are the host party for only 2 % of all CDM projects (Gavaldao et al. 2012). By comparison, China, India, Brazil, and South Korea collectively host approximately 75 % of CDM projects.

To address this disparity, the CDM was encouraged by Kyoto subscribers to develop methodologies where suppressed demand could be applied "prioritizing those that are more applicable to least developed countries, small island developing states, African countries and countries underrepresented in the clean development mechanism." (CDM 2012). These priorities reflect the sustainable development objective of

Kyoto, and thus need not be based on actual reductions in carbon loads. The CDM published guidelines that clarify that suppressed demand is applicable "when a minimum service level to meet basic human needs … was unavailable to the end user of the service prior to the implementation of the project activity." (CDM 2012).

The GS also authorizes the use of suppressed demand in calculating credits, either by using the CDM technologies (which the GS expressly authorizes) or its own TPDDTEC methodology. That methodology includes a definition of suppressed demand similar to that found in the CDM (GSF 2011).

### 9.1.5   Suppressed Demand and HWT

As noted above, suppressed demand can be applied in both CDM water treatment methodologies and to water treatment projects under the GS TPDDTEC methodology. Under the CDM, in settings with <60 % improved water coverage (Case 1), credits are available for the full population without access to improved water, with no need to show the actual proportion that are treating their water by boiling. In all other settings (Case 2), credits are only available for the proportion of the unimproved water households (CDM 2013b). That proportion is calculated by estimating the percentage of households that boil (as opposed to some other treatment method) and then projecting that over those that do not practice any form of HWT. Thus, by way of example, assume that in a population of 100 households, 90 are relying on unimproved water supplies and that of these, 20 are boiling, 10 are chlorinating and 60 are not treating their water. The 2:1 ratio of boiling to chlorinating would be applied to all 90, yielding potential credits for 60. In either case, calculations of emission reduction start with an artificially higher level of emissions (CDM 2013b). Under the TPDDTEC methodology in the GS, suppressed demand is applied to water treatment projects similarly to how it is applied in case 2 under AMS-III.AV (GSF 2011).

Suppressed demand exaggerates the actual impact of water treatment programs on GHG emissions. While there is a potential future reduction from discouraging householders from boiling, this is largely hypothetical with little empirical evidence. It is based on two assumptions, neither of which is likely to be true: first, that the recipients of these interventions are going to become wealthier in the future; and second, that as their wealth increases, they will boil water as a means of treatment (Starr 2011, 2012). Thus, carbon credit financing of water treatment should not be justified on the basis of the impact on the GHG emissions. As we have seen, however, the Kyoto Protocol intended carbon credits to also advance development objectives.

From a health and development perspective, the fundamental purpose of suppressed demand—extending the benefits of carbon financing to non-industrialized countries—performs an especially valuable role in the case of water treatment. In its absence, programs would need to target the estimated 1.21 billion people that actually boil their drinking water. Like the broader carbon credit program, this would shift the focus to East Asia where boiling is common, and away from South Asia and Sub-Saharan Africa which comprise the largest disease burden from waterborne pathogens (Rosa and Clasen 2010). Moreover, it may simply substitute a prevalent and largely effective approach for treating water (boiling) with a less effective alternative approach, reducing the opportunity for genuine health gains.

### 9.1.6   Documenting Product Use and Water Quality

Credits are verified and issued periodically based on the difference between the baseline emissions and the project emissions for the reporting period. Among the key parameters contained in the formula for calculating credits are number of devices used and usage. In this respect, these programs contain incentives that the program proponent not only secure widespread coverage of the technology, but also their actual use—an obvious but often neglected factor in traditional approaches to implementing environmental health interventions.

For water treatment technologies, the formula estimates the emission reductions based on the volume of water treated at baseline versus under the project. The baseline level is calculated by multiplying the "safe water" consumption of end users observed in the project scenario by the amount of fuel required to boil it. The total safe water consumed in the project scenario is the amount of "safe water" supplied by the project technology and consumed in the project scenario, plus the amount (if any) of raw water boiled after introducing the project technology.

Significantly, references to "safe water" are used in the formula only to distinguish treated water and water from improved sources from "raw" untreated water; no standard is established for its microbiological or chemical parameters. Annex 3 of the GS methodology describes the need for water testing and a justifiable testing procedure, but also allows the project scenario results to be established simply by showing evidence that they are being maintained and used correctly in accordance with manufacturers' instructions.

### 9.1.7   Potential Advantages of Carbon Credit
### Financing of HWT

Despite its limited potential for actually reducing GHG emissions in most cases, carbon financing of HWT is consistent with the objectives of the Kyoto Protocol if it genuinely advances important development objectives in low-income countries. Unlike improvements in water quantity and access, however, which not only advance health but also improve livelihoods in other ways (time savings, food production, etc.), promotion of HWT can be justified only if it actually creates the conditions necessary to reduce waterborne disease (32). As noted above, this requires that it reach a vulnerable population and be used correctly and consistently on a long-term basis (Clasen et al. 2009).

In fact, carbon credit financing may be one of the few options for scaling up HWT in a manner that addresses both of these criteria. First, as described above, credits are maximized in settings where householders are relying on unimproved water supplies, encouraging implementers to target the most vulnerable. Existing HWT projects are based in low-income countries in Africa, Asia and South America. A program in Rwanda focuses solely on the poorest 30 % of the population. This is in contrast, for example, to commercially-driven filter programs or boiling which are taken up mainly by the least poor (Rosa and Clasen 2010). While some still argue that user contributions and cost recovery are necessary to ensure uptake and sustainability of public health interventions, experience from insecticide treated bed nets ultimately showed that only mass free distribution yielded the level of coverage necessary to achieve measurable reductions in malaria (Cohen and Dupas 2006).

Second, credits are only available where the implementer documents use. In this respect, the program encourages implementers not only to deliver inputs (coverage), but to demonstrate outputs (uptake). However, as discussed below, the methods prescribed by the programs to document these parameters are insufficient. More can be done to move the incentives toward health-based outcomes by ensuring that the intervention actually yields improvements in drinking water quality. However, in principle at least, this is a pay-for-performance model that more closely aligns the metrics for program implementation with the conditions necessary to improve health.

Third, credits are available for up to 21 years, as long as the implementer continues to demonstrate coverage and use. Since most of the costs of implementing projects are incurred at the front end by purchasing hardware and paying for its delivery, it can take several years to achieve breakeven. As the ongoing costs for maintaining the project may be comparatively small, especially for durables such as filters, implementers are incentivized to maintain high levels of use even after the sunk costs are covered.

In these respects, carbon financing creates a set of incentives for HWT that are closely aligned with the antecedents for achieving health gains: targeted coverage, consistent use, and sustained use. At the same time, in recognition of the minor impact these programs have on GHG emissions, promoters may wish to refrain from characterizing their programs as trading "carbon for water" and simply refer to them as being supported by "carbon financing".

### 9.1.8  Recommended Changes

Despite these favorable aspects of the program, there are shortcomings in the existing rules governing carbon credit financing of HWT that could reduce its potential for a positive impact on health. Since the use of carbon credits to finance HWT in low-income countries has limited potential of actually reducing GHG emissions, it is essential that the program requirements go as far as possible to create the conditions necessary to prevent disease in order to meet its policy goals.

In respect of water treatment, those conditions are clear: that a population otherwise at risk now have sustainable access to safe drinking water. This goal corresponds with the MDG for water, though the manner in which it has been measured diluted the target to "improved water supplies" instead of "safe drinking water". While the ultimate goal is a health impact, holding implementers responsible for demonstrating a health gain goes too far: safe drinking water is a necessary condition to human health but it will not prevent all potential exposures to microbial pathogens from other pathways, particularly in hygiene challenged settings.

Nevertheless, two changes in the existing carbon credit financing mechanisms for water treatment will help advance the objective of ensuring that these programs deliver safe drinking water. First, the microbiological performance standards should require the technologies to meet the WHO's "highly protective" guideline. Second, the documentary requirements should obligate the implementer to provide independent evidence of actual use. These changes will help ensure the credibility of carbon credit financing for water treatment, and by doing so maintain the value of the credits.

## 9.1.9  Documenting Use and Water Quality Improvement

Carbon credit financing creates incentives that encourage continued use and improved water quality. These go beyond traditional input-based performance measures that rely solely on coverage. Nevertheless, the current procedures are inadequate to ensure these antecedents for eliminating exposure to waterborne pathogens are being fully realized.

Overall, the current methodologies rely heavily on records by program promoters. While promoters should be qualified to provide reliable records on units sold or distributed, conducting surveys to establish use—especially correct, consistent and sustained use, the most critical conditions for preventing disease—is not within their normal expertise. Moreover, while the current CDM and GS methodologies express the need for random sampling, adequate sample sizes and an approach that captures seasonal variations, they do not contain sufficient detail to ensure that the results are methodologically rigorous. With respect to surveys used, for example, the GS methodology prescribes a minimum sample size of just 100 households, even though the programs themselves can cover hundreds of thousands of households. Further, only a majority of those surveys need be conducted on site with observations by the interviewer; the balance can be conducted by phone. From a health perspective, observing or recording the presence of water treatment hardware is not sufficient; evidence shows that even among self-reported users, HWT practices are often exaggerated and ineffective (Rosa et al. 2014). Moreover, children and others still tend to drink untreated water even if they have a filter or other treatment method (Boisson et al. 2010).

Rigorously measuring use on these parameters is not comparable to providing sales or distribution records to document coverage. It requires trained and adequately equipped enumerators following a prescribed protocol with an adequate sampling strategy to ensure internal and external validity. In most cases, this should not be done by the program promoter but by an experienced and independent third party. At the same time, this is not unduly onerous. The WHO has recently led an effort to develop and validate metrics for monitoring and evaluating HWTS programs that includes measures for assessing correct, consistent use (WHO 2012). Instrumented monitoring can also be used to verify the self-reported use (Thomas et al. 2013).

In the end, however, the objective of these programs should be not just use; it should be "effective use"—use that results in the drinking water actually being safe. The best way of assessing such use is not to rely on observations or reports, but to actually test the water. Having recognized the shortcomings of relying on indicators and self-reports, the WHO/UNICEF Joint Monitoring Committee on Water and Sanitation are recommending such household level water testing in national level surveys designed to assess access to safe drinking water (2). The use parameter for purposes of determining carbon credits should require that at an unannounced visit (i) the female head of household actually reports that water is used for drinking, especially by young children, and (ii) a sample of the water is shown by accepted methods to conform with WHO (or higher local) quality standards for microbial contamination.

## 9.2  Conclusion

Carbon financing could make a significant contribution to scaling up HWT among vulnerable populations. Suppressed demand makes the programs feasible in South Asia and Sub-Saharan Africa where they can reach those most at risk. However, because these programs will have little impact on GHG emissions, they must clearly advance the development objective of providing sustainable access to safe drinking water. This objective requires that qualifying programs meet conditions necessary for HWT to contribute to public health: reaching vulnerable populations with microbiologically effective solutions and ensuring that they are used correctly and consistently. Changes to the program provisions that require the technologies to provide complete microbiological protection and be used by the target population in a manner that actually provides safe drinking water will help ensure that the credits are associated with conditions likely to yield genuine health gains.

## References

Arnoldus M, Bymolt R (2011) Demystifying carbon markets: a guide to developing carbon credit projects. KIT Publishing, Amsterdam

Boisson S, Kiyombo M, Sthreshley L, Tumba S, Makambo J, Clasen T (2010) Field assessment of a novel household-based water filtration device: a randomised, placebo-controlled trial in the Democratic Republic of Congo. PLoS One 5(9):e12613. doi:10.1371/journal.pone.0012613

CDM (2012) Guidelines of the consideration of suppressed demand in CDM methodologies. CDM Executive Board, Bonn

CDM (2013a) Annual report of the executive board of the clean development mechanism to the conference of the parties serving as the meeting of the parties to the Kyoto protocol. UNFCCC, New York

CDM (2013b) CDM Executive Board Methodology Panel, AMS-III.AV. Low greenhouse gas emitting safe drinking water production systems, vol 4. New York

Clasen T, Naranjo J, Frauchiger D, Gerba C (2009) Laboratory assessment of a gravity-fed ultra-filtration water treatment device designed for household use in low-income settings. Am J Trop Med Hyg 80(5):819–823

CMW Carbon Market Watch, Boosting CDM projects in LDCs: an introduction to suppressed demand. http://carbonmarketwatch.org/boosting-cdm-projects-in-ldcs-an-introduction-to-suppressed-demand-newsletter-16/

CMW Carbon Market Watch, Suppressed demand in the CDM. http://carbonmarketwatch.org/category/additionality-and-baselines/suppressed-demand/

Cohen J, Dupas P (2006) Free distribution or cost-sharing? Evidence from a malaria prevention experiment in Kenya. Jameel Poverty Action Lab, Cambridge

Gavaldao M, Battye W, Grapeloup M, Francois Y (2012) Suppressed demand and the carbon markets: Does development have to become dirty before it qualifies to become clean? Field Actions Sci Rep 7(Special Issue).

GSF (2011) Technologies and practices to displace decentralized thermal energy consumption. The Gold Standard, Geneva

The Gold Standard Foundation, Rule Update: requirements for the quality standards to be met by safe water supply projects; http://www.goldstandard.org/wp-content/uploads/2011/09/Rule-Update-Requirements-for-treated-water-quality.pdf

Held RB, Zhang Q (2013) Quantification of human and embodied energy of improved water provided by source and household interventions. J Clean Prod 60:83–92

Kouchakji K, 10 years of gold standard

Peters-Stanley M, Yin D (2013) Maneuvering the mosaic: the state of the voluntary carbon markets 2013. Ecosystem Marketplace, Washington, DC

Rosa G, Clasen T (2010) Estimating the scope of household water treatment in low- and medium-income countries. Am J Trop Med Hyg 82(2):289–300. doi:10.4269/ajtmh.2010.09-0382

Rosa G, Majorin F, Boisson S, Barstow C, Johnson M, Kirby M, Ngabo F, Thomas E, Clasen T (2014) Assessing the impact of water filters and improved cook stoves on drinking water quality and household air pollution: a randomised controlled trial in Rwanda. PLoS One 9(3):e91011. doi:10.1371/journal.pone.0091011

Starr K (2011) Thirty million dollars, a little bit of carbon, and a lot of hot air. Stanford Social Innovation Review Blog, Palo Alto

Starr K (2012) Another look at "Carbon for Water" in Western Kenya. In Stanford Social Innovation Review Blog, http://www.ssireview.org/blog/entry/another_look_at_carbon_for_water_in_western_kenya

Thomas E, Barstow C, Rosa G, Majorin F, Clasen T (2013) Use of remotely reporting electronic sensors for assessing use of water filters and cookstoves in Rwanda. Environ Sci Tech 47:13602–13610

UN (1998) Kyoto protocol to the united nations framework convention on climate change. United Nations, New York

UNFCCC (2006) Decision 3/CMP.1. Report of the conference of the parties serving as the meeting of the parties to the Kyoto Protocol on its first session. Montreal, November 28–December 10, 2005

UNFCCC (2012a) Benefits of the clean development mechanism 2012. Bonn

UNFCCC (2012b) CDM methodology booklet, 4th edn. Bonn

UNICEF, WHO (2013) Progress on sanitation and drinking-water 2013 update. WHO/UNICEF, Geneva

WHO (2011) Evaluating household water treatment options: health based targets and microbiological performance specifications. WHO, Geneva

WHO (2012) A toolkit for monitoring and evaluating household water treatment and safe storage programmes. World Health Organization, Geneva

Williams M, Murthy S (2013) Reconciling the carbon market and the human right to water: the role of suppressed demand under clean development mechanism and the gold standard. Environ Law 43:517–562

World Bank $CO_2$ emission (metric tons per capita). http://data.worldbank.org/indicator/EN.ATM.CO2E.PC. Washington, DC

# Chapter 10
# HAPIT, the Household Air Pollution Intervention Tool, to Evaluate the Health Benefits and Cost-Effectiveness of Clean Cooking Interventions

Ajay Pillarisetti, Sumi Mehta, and Kirk R. Smith

**Abstract** There is a growing focus on interventions seeking to reduce the burden of disease associated with household air pollution. HAPIT provides policy-makers and program implementers an easy-to-use tool by which to compare the relative merits of programs both within and between countries, helping assist with optimization of limited resources. Although a number of uncertainties remain, HAPIT represents the 'state of the science' and relies on the best available knowledge – and is built to easily integrate new knowledge and findings to better hone estimates.

**Keywords** HAPIT • Cookstoves • Household air pollution • ADALY • Cost effectiveness

## 10.1 Introduction

Globally, approximately 40 % of the world's population relies on solid fuel combustion for cooking (Bonjour et al. 2013). The household air pollution (HAP) resulting from the use of these fuels (including wood, dung, coal, and crop residues) results in approximately four million premature deaths yearly (Smith et al. 2014; Lozano et al. 2012) and 108 million lost disability-adjusted life years (DALYs) in low and medium income countries (LMICs). This burden comes from HAP's impact on a range of diseases, including chronic obstructive pulmonary disease (COPD), ischemic heart disease (IHD), stroke, and lung cancer (LC) in adults and acute lower respiratory

A. Pillarisetti (✉) • K.R. Smith
Division of Environmental Health Sciences, School of Public Health,
University of California, Berkeley, University Hall, Berkeley, CA 94720-7360, USA
e-mail: ajaypillarisetti@gmail.com

S. Mehta
Global Alliance for Clean Cookstoves,
1750 Pennsylvania Ave NW, Washington, DC 20006, USA

© Springer International Publishing Switzerland 2016
E.A. Thomas (ed.), *Broken Pumps and Promises*,
DOI 10.1007/978-3-319-28643-3_10

infection (ALRI) in children. In response to this large health burden, international organizations and governments – recently spearheaded in part by the Global Alliance for Clean Cookstoves – have focused on efforts to provide reliable clean cooking technologies to solid fuel users. Deployed interventions span a range of technologies, including simple "improved" chimney stoves (Singh et al. 2012; Smith et al. 2010), 'rocket' stoves (Rosa et al. 2014), advanced cookstoves with fan-assisted combustion (Sambandam et al. 2015; Pillarisetti et al. 2014), as well as clean fuel (including LPG, natural gas, biogas, ethanol and electricity) interventions (Van Vliet et al. 2013; Neupane et al. 2015).

Selecting an intervention requires balancing a number of competing priorities, including the cost of the intervention; its effectiveness, as proven in the lab and pilot field studies; its cultural acceptability, attractiveness, and ability to meet local cooking needs, and its inherent characteristics, like the need for fuel processing, the intervention's durability, and power constraints. One way to frame this selection process is at the scale of a large program, with consideration of its potential to improve quality of life and avoid ill-health not only in absolute terms, but also in relation to resources spent on its deployment and evaluation.

To actually measure the broad range of changes in health from a change in the HAP due to an intervention would require large, complicated, expensive, long-term field studies, particularly as the prevalence of most of the chronic diseases known to be exacerbated by HAP (COPD, IHD, LC, stroke) take many years to develop but also many years to decline with reductions in exposure. There is nevertheless a need for methods to credibly estimate the likely degree of ill-health that could be avoided by an intervention using the best available scientific evidence from epidemiological studies that could be expected from an intervention.

In this chapter, we describe the development of and methodology used in the Household Air Pollution Intervention Tool (HAPIT), an internet-based platform[1] to evaluate and compare health benefits achievable through reduced exposures to fine particulate matter ($PM_{2.5}$) resulting from implementation of fuel and/or stove interventions. It can be tailored to the conditions in each of many dozens of LMICs to give organizations contemplating interventions a rough, but credible, estimate of the comparable health benefits that could be accrued through each scenario.

The idea behind HAPIT is not to provide research-quality evidence of health benefits for all possible situations, which would take many years and involve costs and expertise that is well beyond that possible as part of most planned interventions. Rather, it aims to provide "good enough" evidence based on the best available health effects information linked to air pollution exposures. There is a long tradition of using such risk assessment techniques to evaluate environmental health hazards not only in air pollution (EPA) but from interventions to reduce water pollutants, radiation, toxic chemicals, and so on.

Evaluations of projects to reduce another important environmental health risk also benefit from such tools. Interventions to mitigate climate change use $CO_2$-equivalent metrics to estimate their benefits. They are not required to actually show

---

[1] HAPIT can be accessed at http://hapit.shinyapps.io/HAPIT

an impact on climate change, which would take sophisticated studies lasting many years, but rely on links established by the best current science between emissions of greenhouse gases and changes to climate. These come from complex climate models informed by measurements and that are evolving over time. Just so with HAPIT, which relies on the best intermediate variable between HAP and health, exposure to $PM_{2.5}$. Exposure is closely linked to the intervention in one direction and to health impacts in the other via complex published models based on major reviews of health studies in real populations, which, like climate change models, evolve over time.

HAPIT outputs can be shared with policy makers in order to raise awareness about the potential public health implications of the program at a national level, inform them about the relative health benefits expected by scaling up available interventions, and provide information on the relative costs of scaling up different intervention options. As such, there is a clear role for such a tool to inform health policy makers in the implementation of the World Health Organization's Indoor Air Quality Guidelines focused on household fuel combustion. Beyond the health sector, this tool can be used by clean cooking implementers (1) to help both design better interventions (how clean do interventions need to be to achieve health benefits) and (2) to potentially help raise funds to implement dissemination projects through results-based financing.

HAPIT estimates both averted DALYs (aDALYs) and averted premature deaths and calculates a simple cost-effectiveness metric based on the World Health Organization's Choosing Interventions that are Cost-Effective (WHO-CHOICE) framework. For illustration, we demonstrate use of HAPIT to evaluate a chimney stove intervention deployed as part of the RESPIRE randomized controlled trial and an LPG intervention, both in the Western Highlands of Guatemala. Finally, we conclude with a discussion of the methodological and conceptual issues raised by HAPIT in the context of broader health and sociopolitical concerns and introduce the potential for results-based financing based on averted DALYs.

## 10.2  Methods

HAPIT relies (1) on up-to-date national background health data and (2) on the methods and databases developed as part of the Comparative Risk Assessment (CRA), a component of the 2010 Global Burden of Disease (GBD 2010). HAPIT utilizes exposure-response information for each of the major disease categories attributable to particulate air pollution and 2010 background demographic, energy, and economic conditions for the 57 countries in which solid fuels are the primary cooking fuel for 50 % or more of homes (Bonjour et al. 2013). HAPIT additionally includes a number of countries in which household energy disseminations are underway or planned but who have less than 50 % solid fuel use nationally. All data are for year 2010, the most recent year for which country-level data are currently available from GBD. Figure 10.1 visually depicts HAPIT inputs and methods.

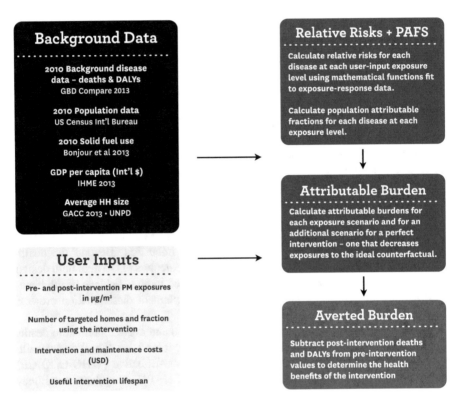

**Fig. 10.1** A conceptual diagram of the inputs, outputs, and methods used to estimate averted ill-health using HAPIT

## 10.2.1  Background Data Used by HAPIT

All background disease information employed in HAPIT was downloaded from the Institute for Health Metrics and Evaluation's (IHME) GBD 2010 Country Databases. The deaths and DALYs from lung cancer includes the GBD 2010 estimates of trachea, bronchus, and lung cancers. Cardiovascular diseases are broken down into two categories – Ischemic Heart Disease (IHD) and Ischemic & Other Hemorrhagic Strokes (Stroke). HAPIT calculates deaths and DALYs due to ALRI only among the population of 0–4. Average household sizes were extracted from the Global Alliance for Clean Cookstoves' Data and Statistics website (GACC). Population data were extracted from the US Census International Bureau (USCB 2015) and the UN's World Urbanization Project (UNDESA 2014).

Cost-effectiveness is determined by comparing the expected annual cost of the intervention per averted DALY (described below) in USD to the gross domestic product per capita (GDP PC, USD). WHO-CHOICE advises that interventions costing less than the GDP/capita are very cost-effective, those costing one to three times the GDP/capita are cost-effective, and those costing more than three times the GDP/capita are not cost-effective (Evans et al. 2006).

HAPIT estimates program cost-effectiveness using a financial accounting approach. In doing so, it (1) does not take into account changes in household costs due to medical expenditure or the time or money spent acquiring fuel and it (2) assumes that programs are covering the cost of fuel-based interventions (such as monthly LPG costs per household). For custom scenarios, users can adjust the per-household maintenance or fuel cost based on the characteristics of their programs to take into account these parameters. For example, the total financial outlay of the intervention program may decrease if households pay for a portion of the fuel or intervention cost up-front or over time.

## 10.2.2   User Inputs

HAPIT users are able to input (1) pre- and post-intervention population average exposures to $PM_{2.5}$ in $\mu g/m^3$, based on measurements performed in the target communities, and the standard deviation of those measurements; (2) the number of households targeted by the intervention; (3) the average percentage of the population using the intervention throughout the intervention's useful lifetime; (4) the cost per intervention to the program in current US Dollars (USD); and (5) the yearly maintenance cost (including fuel costs) per household in current USD. For users with limited knowledge to inform these inputs, default values are available for all of the above.

Users are strongly urged to address the following concerns prior to scaling up an intervention:

- Intervention Effectiveness: selected interventions should have the ability, under ideal conditions, to reduce emissions of health damaging pollutants to acceptable levels, as assessed in the laboratory (Jetter et al. 2012; Jetter and Kariher 2009). Interventions that perform poorly in the laboratory are unlikely to perform well in the field.
- Intervention Acceptability: interventions should be fully vetted by community members to ensure appropriateness for local cooking and otherwise to suit local needs.
- Exposure Reduction: interventions should result in a demonstrable and significant reduction in $PM_{2.5}$ population exposures in pilot work in the community of interest, or one like it.
- Sustained Intervention Usage: declining usage of the intervention over time may indicate reversion to traditional cooking methods and an elimination of any meaningful exposure reductions. Interventions should be used regularly and should, ideally, displace use of the more polluting traditional stove.

Because HAPIT relies on measured exposures to estimate averted ill-health, we briefly clarify the distinction between (a) emissions, (b) concentrations, and (c) exposures in the context of household air pollution studies:

(a) <u>Emissions</u> refers to the rate of release of a pollutant per unit time or per unit of fuel; emissions measurements are often taken 'directly' from the combustion source and can be performed in the laboratory or the field. Although emission measurements can be conducted over an entire day, it is most common to conduct them in conjunction with one cooking cycle, either typical to the area if done in the field or with a standard cooking cycle if done in the lab.

(b) <u>Concentrations</u> are a result of emissions and various room conditions, like ventilation rates, and processes, like deposition and exfiltration. Concentrations are often measured in microenvironments – for instance, in the kitchen and in the living room – but do not directly take into account the presence of people. Because it is difficult to simulate real world situations, reliable concentration measurements normally are measured in households themselves. Commonly, for example, kitchen air pollution (KAP) measurements are made by placing a monitor on the wall of the kitchen for 24 h.

(c) <u>Exposures</u> are complex, spatiotemporal relationships between individuals and the concentrations of pollutants in their vicinity. A population exposure thus depends on the concentration of pollutant in an area, the number of people in the area, and the time spent in that given area. Similarly, an individual's daily exposure is impacted by the variety of sources they experience in the spaces they inhabit for varying lengths of time throughout the day. For example, high concentrations of pollutants in a smoky kitchen do not necessarily result in high exposures; if the cook spends most of her time outside of the kitchen, her average exposure may not be as high as a concentration may predict. Exposure involves contact between humans and pollution. Because of the nearly universal diurnal pattern of human activity, exposure monitoring is best done for at least 24 h or in multiples of 24 h (48, 72, etc).

Data on lab-based emissions, although fewer than desirable, are increasingly publically available (catalog.cleancookstoves.org). In contrast, the availability of exposure data across a range of geographies, fuel and stove combinations, and cooking practices remains limited, especially for the most promising (based on lab performance) stoves and fuels. Moreover, given the complexity of exposure characterization and the paucity of available data linking exposure and emissions, it is not currently possible to reliably estimate exposures based on lab-based emissions data without extensive measurements followed by modeling at the local level. Default exposures in HAPIT are based on the currently available literature and informed largely by global modeling of HAP exposures (Balakrishnan et al. 2013; Smith et al. 2014).

### 10.2.3    Integrated Exposure-Response Functions

Estimating the burden of disease attributable to all types of air pollution – including household air pollution (HAP) – during the 2010 Global Burden of Disease required elaboration of integrated exposure-response (IER) relationships (Burnett et al. 2014)

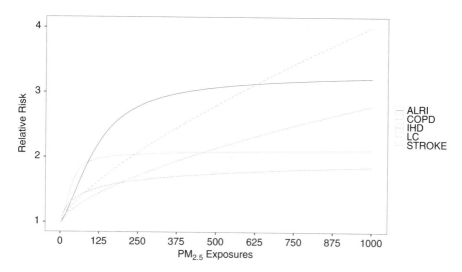

**Fig. 10.2** Integrated Exposure Response (IER) curves relating Exposure to PM$_{2.5}$ to health end-points associated with exposure to air pollution, including ischemic heart disease (IHD), stroke, chronic obstructive pulmonary disease (COPD), and lung cancer (LC) in adults and acute lower respiratory infection (ALRI) in children (See Fig. 10.4 for an elaboration of uncertainties around the IERs)

that relate PM$_{2.5}$ exposures to risk for a number of health endpoints. The IERs leverage epidemiological evidence from a wide range of PM$_{2.5}$ exposures spanning multiple orders of magnitude (ambient air pollution, active and secondhand tobacco smoke, and household air pollution) and result in supra-linear exposure-response curves (Fig. 10.2).

In Burnett et al. the parameterization of the IERs took a common form:

$$RR_{IER}(z) = 1 + \alpha \left\{ 1 - \exp\left[ -\gamma \left( z - z_{cf} \right)^{\delta} \right] \right\}$$

(10.1)

where z is exposure to PM$_{2.5}$ in μg/m$^3$, $z_{cf}$ is the counterfactual exposure to PM$_{2.5}$ in μg/m$^3$, and where α, γ, and δ are model parameters. In initial versions of HAPIT (version 1 and 2), Eureqa (Nutonian, Inc.) was used to fit a line to a table of central relative risk estimates (and lower and upper confidence bounds) provided by Burnett et al. for concentrations ranging from 0 to 1000 μg/m$^3$. In version 3 of HAPIT, we utilize data released by the Institute of Health Metrics and Evaluation (IHME) to create a lookup table of relative risks, using methods similar to those reported elsewhere (Apte et al. 2015). For each health endpoint – and for twelve age-categories for IHD and Stroke – 1000 values of $z_{cf}$, α, γ, and δ were provided (IHME 2010). We calculated the lower bound (5th percentile), central (mean), and upper bound (95th percentile) relative risk estimates from the distribution of provided values for each health endpoint, age-category, and for exposures ranging from 1 to 1000 μg/m$^3$ in discrete 1 μg/m$^3$ steps. For concentrations less than the counterfactual concentration of 7.3 μg/m$^3$, the relative risk was fixed at 1 – an indication of no difference in risk.

### 10.2.4   Evaluating Averted Ill-Health

HAPIT generates 1000 pairs of pre- and post-intervention exposures by sampling from a lognormal distribution reconstructed from the user input mean exposure and measurement standard deviation. For each pair of exposures, HAPIT identifies the corresponding relative risks from the look-up table. The population attributable fraction (Eq. 10.2) is then calculated as follows:

$$PAF = \frac{SFU(RR-1)}{SFU(RR-1)+1} \tag{10.2}$$

where SFU refers to the percent of the population using solid fuels and RR refers to the relative risk calculated using the IERs. The approach utilized is based on methods developed by the GBD and others (Smith and Haigler 2008; WHO 2004), but adapted to take advantage of the continuous IERs.

To estimate changes in deaths and DALYs attributable to an intervention ($AB_{int}$), we subtracted the PAF after the intervention ($PAF_{post-intervention}$) from the PAF prior to the intervention ($PAF_{pre-intervention}$) and multiplied by the user input usage fraction; the underlying disease burden ($B_{endpoint}$) for a specific country, health endpoint, and age-group as follows; and the percentage of solid-fuel use in the target population:

$$AB_{int} = \left(PAF_{pre-intervention} - PAF_{post-intervention}\right) \times B_{endpoint} \times Use_{fraction} \times SFU_{fraction} \tag{10.3}$$

Averted burdens are calculated for all combinations of the lower, central, and upper relative risk estimates and the central background disease rate estimates for each of the 1000 exposure pairs. HAPIT outputs the following:

(a) the mean averted deaths and DALYs – the mean of the 1000 attributable burdens calculated using the central relative risk
(b) the minimum averted deaths and DALYs – the mean of the 1000 attributable burdens calculated using the lower bounds of the IERs
(c) the maximum averted deaths and DALYs – the mean of the 1000 attributable burdens calculated using the upper bounds of the IERs
(d) the maximum avertable deaths and DALYs – the burden that could be averted by going from the pre-intervention exposure to the counterfactual, assuming 100 % stove usage

HAPIT assumes that all deaths and DALYs due to ALRI are accrued instantaneously upon implementation of the intervention. For chronic diseases in adults (COPD, stroke, IHD, and lung cancer), HAPIT utilizes the 20-year distributed cessation lag model of the United States Environmental Protection Agency (US EPA), a step function for estimating the accrual of benefits due to changes in exposure to air pollution (Fig. 10.3). The EPA model assumes that 30 % of mortality reductions occur in the first year, 50 % are distributed evenly in years two through

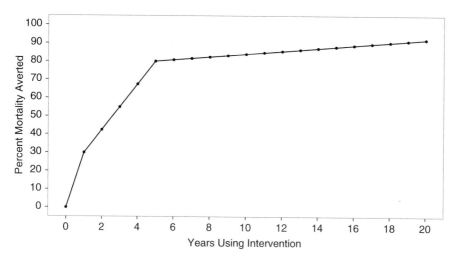

**Fig. 10.3** Visual representation of the EPA 20-year cessation lag function. The cessation lag function as outlined by the US EPA is used to adjust downward the attribution of averted DALYs and Deaths from chronic disease due to reduced PM2.5 exposures resulting from an HAP intervention

five, and the remaining 20 % are distributed evenly in years six through twenty (EPA 2004). At the end of the intervention's lifetime, we assume that benefits for children from reduced ALRI cease; an additional 75 % of a full benefit-year accrue for chronic diseases.

HAPIT limits an intervention's useful lifetime to, at a maximum, 5 years. This is due to two issues. First, because attributable burden calculations rely on up-to-date background disease information, extending beyond 5 years unrealistically assumes no change in background disease rates. Second, evidence from the field indicates that many current interventions do not have a useful life beyond 2 or 3 years (Pillarisetti et al. 2014; Hill et al. 2015) at most.

## 10.2.5  Averted Disability Adjusted Life-Years (DALYs)

While HAPIT outputs averted deaths, a perhaps more interesting and useful output is that of averted DALYs. The DALY is a combined metric of mortality and morbidity that measures the gap between the 'ideal' and the experienced health states of a population. DALYs are composed of two parts: years of life lost (YLLs) to premature death and years lived with disability (YLDs) weighted by the severity of the condition experienced. Fundamentally, the DALY seeks to put death and disability from all diseases on an equal footing for all individuals of the same age in the world, irrespective of social class, country of origin, socioeconomic status, occupation, or other characteristic (Mathers et al. 2006). GBD 2010 used a life expectancy at birth of 86 years to calculate YLLs and, unlike previous GBD undertakings, removed all discounting and age-weighting (Murray et al. 2012). The calculation of disability weights was updated to take into account global heterogeneity in perception of the severity of various conditions and was utilized revised methods by which surveys were translated into severity weights. While a number of concerns about the use of the DALY remain (Voigt and King 2014), to date no other combined metric of morbidity and mortality has been as thoroughly described and used in global health literature. Use of the DALY allows simple comparison of cost-effectiveness across sectors and potential interventions and is commonly used in the global health literature.

### 10.2.6  Implementation

The basic calculations for HAPIT are implemented in R 3.1 (R Foundation) and utilize Shiny, a framework enabling sharing of interactive R code over the internet (Chang et al. 2015). HAPIT is currently hosted by RStudio for a nominal monthly fee. Figures are generated using ggplot2 (Wickham 2009).

## 10.3  HAPIT in Use – Hypothetical Chimney Stove and Liquefied Petroleum Gas Interventions in the Western Highlands of Guatemala

### 10.3.1  Background

As an illustration of the use of HAPIT, we adapt findings from the Randomized Exposure Study of Pollution Indoors and Respiratory Effects (RESPIRE), a randomized control trial (RCT) that assessed the impact of reduced emissions from a chimney stove on childhood pneumonia (Smith et al. 2011), and subsequent studies in the region (Smith et al. 2010; McCracken et al. 2007, 2011; Northcross et al. 2010). The study design has been described extensively elsewhere (Bruce et al. 2007; Smith-Sivertsen et al. 2009). Briefly, it took place in the Western Highlands of Guatemala between October 2002 and December 2004. Most study homes were located between 2000 and 3000 m above sea level and used wood as their primary cooking fuel. Five hundred and eighteen households contributed to the final dataset, with approximately half receiving a chimney stove and the other half cooking using the traditional open fire. Across Guatemala, 64 % of households rely on solid fuel for cooking. The GDP per capita in Guatemala is approximately 5000 USD.

## 10.3.2   Inputs

While carbon monoxide (CO) exposures were the primary exposure measurement collected during RESPIRE, $PM_{2.5}$ exposures were also assessed at various points throughout and after the primary RESPIRE trial, including as described in McCracken et al (2007, 2011). For this analysis, we assume any new chimney stove implemented in the region would perform similarly to findings during those assessments; that is, we expect to see adult exposure reductions to $PM_{2.5}$ from 264 µg/m³ (SD = 297) when using the traditional stove to 102 µg/m³ (SD = 130) when using the intervention chimney stove.[2] For children, we use the ratio of child to mother exposures to carbon monoxide to scale exposure reductions appropriately. Because of the rich data available on these exposures, we are able to estimate mother to child ratios for both the pre-intervention and post-intervention periods. During the pre-intervention period, child exposures are ~45 % of the mother's exposure; in the post-intervention period, child exposures are ~54 % of the mother's exposure. Accordingly, for children, the pre- and post-intervention $PM_{2.5}$ exposures are 119 µg/m³ (SD = 133) and 55 µg/m³ (SD = 70), respectively.

We additionally assume the intervention will reach 25,000 households, be used consistently by 90 % of households as previously reported (Ruiz-Mercado et al. 2013), have a 5-year lifespan, cost 200 USD per stove, and have a maintenance cost of 5 dollars per year per stove. For comparison, we will also consider an LPG intervention that reduces exposures of both mothers and children to the level of ambient pollution in these communities of 30 µg/m³ (SD = 20),[3] has an identical useful lifespan and fraction of households using the intervention, and costs 75 USD per stove with a fuel cost of 175 USD per year per household. Inputs for both scenarios are summarized in Table 10.1.

## 10.3.3   Findings

Figure 10.4 depicts the simulated exposures before and after distribution of the chimney-stove intervention. The depicted IERs illustrate the non-linear nature of expected health-benefits associated with an exposure reduction. For instance, for adults, the relationship is relatively 'flat' for Stroke and IHD for an exposure

---

[2] Application of HAPIT should ideally include up-to-date personal exposure measurements of $PM_{2.5}$.

[3] The post-intervention concentration in this LPG scenario may seem counter-intuitive: LPG has been shown to very clean and emit almost no particles when operated properly. Why, then, not reduce the exposure to the ideal, 7.3 µg/m³ counterfactual? In this case, we assume some pollution arises from households in the community who may not have transitioned to LPG or from other sources, such as trash burning, power generation, or vehicles, to name a few possibilities. LPG exposure reductions for this example are set to background ambient $PM_{2.5}$ concentrations as measured during RESPIRE.

**Table 10.1** HAPIT inputs for chimney stove and LPG interventions in rural Guatemala

| | Pre-intervention exposure μg/m³ (SD) | | Post-intervention exposure μg/m³ (SD) | | #Homes | Average use % | Stove lifetime (years) | Initial cost USD | Yearly cost USD |
|---|---|---|---|---|---|---|---|---|---|
| | Adults | Kids | Adults | Kids | | | | | |
| Chimney | 264(297) | 118(113) | 102(130) | 55(70) | 25,000 | 90 | 5 | 200 | 5 |
| LPG | | | 30(20) | | | | | 75 | 210 |

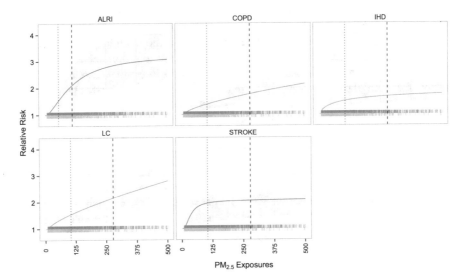

**Fig. 10.4** Integrated exposure-response curves and uncertainty bounds (*lightly shaded*) for each of the major disease categories associated with exposure to HAP. The *dashed vertical line* indicates the pre-intervention exposure; the *dotted vertical line* indicates the post-chimney intervention exposure. The *upper* and *lower tick marks* along the x-axis are the distributions of the simulated pre- and post-intervention exposures, respectively

reduction due to the intervention. For COPD and lung cancer in adults and ALRI in children, the relationship is relatively linear, though the slope varies. For all health endpoints, the uncertainties are large and variable depending on the location on the curve corresponding to a specific exposure.

### 10.3.4 Assumptions

In the above examples, we do not consider the common practice of stove stacking, which would result, most likely, in modified post-intervention exposures. We do not include costs or savings to households, which may include time saved and put towards other productive activities. Additionally, we do not consider dissemination costs or monitoring and evaluation costs, though as mentioned above we do assume that fuel costs are covered by the program. We assume that background disease rates for all of Guatemala are applicable to this region.

Estimates from HAPIT suggest that dissemination of 25,000 chimney stoves – similar to those used during the RESPIRE RCT – with 90 % usage, no stove stacking, and a 5-year lifespan would avert approximately 3270 DALYs (uncertainty bounds 1760–4470) and 65 (uncertainty bounds 35–90) deaths given the exposure reductions modeled above. The majority of the health benefits result from reductions in ALRI in children under 5 (Table 10.2). Figure 10.5a displays the Averted DALYs and Deaths by disease category and the burden remaining for each group. On average, approximately 72 % of the burden remains, though there is heterogeneity between disease categories (range: 62–85 %). When using the least conservative estimate, approximately 62 % of the burden still remains. Similarly, 57 % (range: 57–75 %) of the burden remains if trying to reach 30 $\mu g/m^3$, the level of background ambient pollution in RESPIRE communities.

For an LPG dissemination of 25,000 stoves with 90 % usage, no stove stacking, and a 5-year lifespan, HAPIT estimates approximately 5700 DALYs (uncertainty bounds 3750–6360) and 125 deaths averted (uncertainty bounds 80–160). Figure 10.5b displays the Averted DALYs and Deaths by disease category for an LPG intervention as described. On average, approximately 52 % percent of the burden remains (range 39–69 %). When using the least conservative estimate of the potential impact of an LPG intervention, approximately 39 % of the burden remains. Contrastingly, the ill-health remaining on the table relative to ambient air pollution is only approximately 16 %. This latter would be taking ambient air pollution as the counterfactual, i.e. the minimum achievable by a change within the household itself.

Despite its large unaverted burden, the chimney stove intervention is considered 'very cost effective' across its entire range of potential averted DALYs using the simple WHO-CHOICE rubric (Fig. 10.6a). The LPG stove intervention is also considered very cost-effective, though the range of uncertainty around this estimate is greater than for the chimney stove (Fig. 10.6b), extending into the "cost-effective" range. The LPG intervention is sensitive to price shocks; if the January 2015 price for an LPG cylinder is used (18 USD), the intervention and its uncertainty bounds move entirely into the "cost-effective" category. In these examples, the households may be willing to bear part of the cost of either a chimney stove or LPG stove and/or the monthly cost of the LPG, thus reducing the direct cost to the program itself and impacting cost-effectiveness estimates.

**Table 10.2** HAPIT outputs for chimney stove and LPG stove interventions in Guatemala

| | ALRI | | COPD | | IHD | | Lung cancer | | Stroke | |
|---|---|---|---|---|---|---|---|---|---|---|
| | DALYs (range) | Deaths (range) | DALYs (range) | Deaths (range) | DALYs (range) | Deaths (range) | DALYs (range) | Deaths (range) | DALYs (range) | Deaths (range) |
| Chimney | 2385 (1290–3230) | 30 (15–40) | 240 (180–290) | 7 (5–10) | 380 (240–580) | 20 (10–25) | 80 (50–90) | 3 (2–4) | 250 (95–330) | 12 (5–15) |
| LPG | 3900 (2570–4780) | 45 (30–55) | 390 (290–470) | 12 (8–14) | 730 (540–1150) | 35 (25–55) | 130 (85–150) | 5 (4–6) | 600 (280–660) | 30 (15–30) |

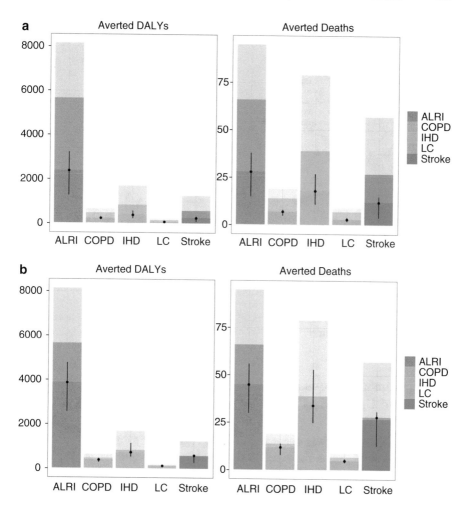

**Fig. 10.5** Averted deaths and DALYs by disease category for (**a**) a chimney stove intervention and (**b**) an LPG stove intervention in Guatemala. The *darkest bars* are the central estimate of averted ill-health; the *lightest bars* are the total burden avertable by the best possible intervention – one that gets down to the counterfactual exposure of 7 μg/m³. The remaining bar represents the burden left by an intervention that gets down to 30 μg/m³, the outdoor ambient level measured during RESPIRE. *Vertical lines* indicate the range of averted ill-health attributable to the intervention modeled by HAPIT

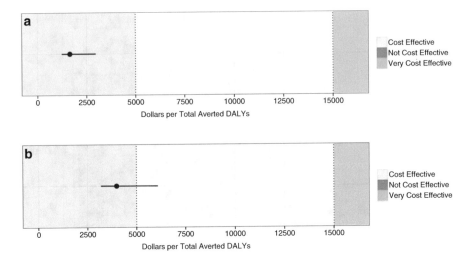

**Fig. 10.6** Dollars per total averted DALYs. The green shading indicates the WHO-CHOICE "very cost-effective category" (< GDP PC per DALY), the yellow shading indicates the "cost-effective" category (between 1 and 3 x GDP PC per DALY) and the red indicates "not cost-effective." The (**a**) *top panel* is for the chimney stove intervention; the (**b**) *bottom panel* is for the LPG intervention. The 2010 GDP PC in Guatemala was approximately 5000 USD

## 10.4    Considerations Arising During the Development and Use of HAPIT

HAPIT provides an easy-to-use, web-based application for assessing the impact of a household air pollution intervention for countries in which there is a significant solid fuel using population. It estimates a range of DALYs and deaths averted by an intervention based on epidemiological methods and using the best available background disease and exposure-response data available. The somewhat simple interface masks significant computational and methodological complexity, and should thus be used with care when making significant policy decisions and considering large interventions with substantial financial and logistical costs.

During the development of HAPIT, a number of methodological and conceptual issues came to the fore. We conclude with a discussion of these issues, of the limitations of HAPIT, and of next steps to further enhance the robustness and reliability of HAPIT-based estimates.

### 10.4.1    Assumptions and Limitations of HAPIT

HAPIT makes a number of assumptions and has a number of limitations. The most prominent follow.

(1) HAPIT assumes that measurements of changes in exposure made over a short period of time are indicative of long-term trends. For results-based financing centered around using averted DALYs and deaths, it will be necessary to perform periodic verification of benefits throughout the period of time financing is sought.

(2) As currently designed, changes in exposure to the cook, upon whom measurements were taken, reflect changes in other household members. The impact on children under the age of 5 is adjusted by the default relationship described above for all scenarios in HAPIT unless an alternate ratio is provided. It is strongly suggested that any alternate ratio be grounded in measurements in the community of interest.

(3) HAPIT assumes background disease and economic characteristics are relatively static. For interventions with a short life-span, this assumption may hold; for long-lived interventions, such as transitioning a community to clean fuels or electricity, HAPIT estimates would need to be revised regularly. In addition, economies of scale are not considered when evaluating cooking interventions costs. Human development indicators may change rapidly depending on social, economic, and political conditions in countries in which HAPIT may be used. These changes can impact the relative merits of a HAP intervention, swaying an intervention from not cost-effective to cost-effective based, for example, on more recent GDP per capita estimates or, for fuel interventions, on fuel costs. For example, the price of LPG in Guatemala has been fairly volatile, varying between 5 USD in 2003 and 18 USD in early 2015 before dropping back down to 10 USD in May of 2015. HAPIT's simplistic cost-estimates do not currently account for monthly or yearly fuel price fluctuations.

(4) HAPIT currently relies on IHME's GBD of disease data, which is, as of now, the most complete and comprehensive burden of disease data available. This completeness comes with the price of some methodological opacity. Continued burden of disease efforts from the World Health Organization and others may result in more rigorous and open model comparison efforts, similar to those seen among climate scientists.

HAPIT highlights the tension between cost-effectiveness and the burden left 'on the table.' As seen in both of the example scenarios above, deployment of an intervention seeking to reduce household air pollution leaves significant ill-health in target communities. This "unaverted" burden poses a quandary to policy makers and health practitioners seeking cost-effective solutions to myriad health problems. The aforementioned chimney stove example is more cost-effective by the admittedly simple form of WHO-CHOICE implemented here; however, it leaves a substantial health burden on the table. The LPG intervention, meanwhile, is less cost-effective, but removes more of this burden from the table. Some may argue that the chimney stove represents an incremental change toward cleaner energy systems; others may counter that leapfrogging attempts at cleanly burning biomass may represent the clearest path forward towards reducing the HAP-related health burden. Rather than make an argument in either direction, we highlight the types of fundamental questions that HAPIT brings to light. These questions are further complicated by considering other health programs – such as a rotavirus vaccine program, the wide-spread deployment of insecticide-impregnated bednets, efforts to improve access to pre-natal care services or a scale-up of water purification devices – side-by-side with HAPIT-based avoided ill-health estimates from clean cooking interventions.

As both interventions leave a significant portion of the burden of the table, we assume that there is some background ambient air pollution – from unclean cooking around intervention homes or from other sources – that contributes to exposures. Controlling this air pollution, by for example ensuring widespread access to clean cooking fuels in a community, could lead to more substantial benefits of an intervention. Put another way, deploying interventions to a larger fraction of homes may have the additional benefit of improving ambient air pollution enough to make an intervention appear more cost-effective. Further research is needed to better understand how much population 'coverage' with and usage of the intervention would be needed to maximize benefits. Finally, our consideration of the burden 'left on the table' explicitly acknowledges that reaching a state of no additional ill-health above the counterfactual would most likely require action to reduce all sources of air pollution – including ambient air pollution from non-cooking sources and pollution released by industries and vehicles, to name a few.

Complications are additionally introduced by an appliance-model of household energy use, in which interventions are used concomitantly with traditional cooking technologies to fully meet the cooking and heating needs of the household – a phenomenon known as stove stacking. As shown in a recent modeling exercise, occasional use of a traditional stove can lead to significant exposures (Johnson and Chiang 2015). HAPIT assumes displacement of the traditional stove for the user-specified percentage of households using the intervention. In homes where stacking occurs, HAPIT may over-predict potential health benefits. Part of this shortcoming is accounted for in the probabilistic approach used, in which 1000 exposures across the distribution of measurements are drawn to estimate averted health impacts. However, the potential impact of stacking to dilute potential exposure reductions should not be dismissed (Pillarisetti et al. 2014; Ruiz-Mercado and Masera 2015).

HAPIT estimates will evolve as GBD-provided background disease information and integrated-exposure response curves change over time. Forthcoming data to be released as part of the 2013 GBD update will undoubtedly alter HAPIT estimates, as it includes a number of revisions to the way air pollution burdens are estimated. Updating HAPIT to account for changes in background disease rates estimated by GBD and for updates to the IER curves is a non-trivial task complicated by the unavailability of programmatic access to GBD data. Furthermore, updates to HAPIT may invalidate results from previous versions of HAPIT.

### 10.4.2    Future Steps

More nuanced probabilistic uncertainty analysis is possible given the wide number of inputs (and corresponding uncertainty bounds) used in HAPIT estimates. Incorporating and propagating these uncertainties throughout the model, however, requires significant computational resources and increases the requisite run time by 10–30 fold. We are evaluating methods to more quickly incorporate these types of uncertainty analyses in HAPIT by utilizing multi-core computing techniques.

An additional and less tractable complexity arises from the model-based uncertainty bounds generated by the IHME modeling of the GBD. As noted elsewhere (Byass 2010; Byass et al. 2013), the uncertainties presented in the GBD 2010 are complex and challenging to interpret and use in further analyses of the type we describe. For some HAPIT parameters, including the IERs and the WHO solid fuel use estimates, more methodological clarity is available, facilitating Monte Carlo and other simulation-based analyses.

### 10.4.3 Including Reductions in Community-Scale Ambient Air Pollution

A well-performing, well-used intervention may result in benefits to households not using the intervention by way of reductions in emissions contributing to ambient air pollution. Accounting for these benefits without a significant measurement campaign is challenging. Measurement may prove feasible; for example, exposure reductions due to reduced ambient pollution can be estimated by (1) by measuring exposures on individuals who did not receive an intervention and (2) by measuring ambient pollution continuously in villages both before and after the deployment of interventions. These measurements are typically expensive, though may be worth pursuing if a program perceives the benefit to be substantial.

HAPIT does not currently have a distinct module to estimate these benefits, though they could be separately estimated in an analogous fashion to those stemming from an intervention. For instance, if measurements indicated that exposures were reduced from 264 $\mu g/m^3$ to 200 $\mu g/m^3$ for an additional 10,000 households, HAPIT could be run using these measurements to estimate the additional averted DALYs and deaths attributable to the intervention's contribution to cleaning up the community airshed as a secondary benefit.

### 10.4.4 Enabling Users to Perform Sub-National or Customized Averted Health Estimates Using Custom-Input Background Disease Data

For some countries – including India, Mexico, Peru, and Nepal – where national statistics may not adequately represent sub-national populations, the ability to customize background disease information may enhance HAPIT's reliability. We are exploring methods by which to incorporate this feature.

## 10.5  Conclusion

There is a growing focus on interventions seeking to reduce the burden of disease associated with household air pollution. HAPIT provides policy-makers and program implementers a relatively easy-to-use tool by which to compare the relative merits of programs both within and between countries, helping assist with optimization of limited resources. Although a number of uncertainties remain, HAPIT represents the 'state of the science' and relies on the best available knowledge – and is built to easily integrate new knowledge and findings to better hone estimates.

HAPIT is freely available for use over the web and can output a summary report to guide later discussions. Like other publically available tools used to assist in resource allocation and policy making decisions (Winfrey et al. 2011; Thompson and Juan 2006), though, it requires a significant understanding of the particulars of the community and country in which an intervention is proposed; confidence in the interventions' ability to reduce exposure to HAP; measurements of exposure to $PM_{2.5}$ before and after an intervention; and significant consideration of optimal ways to deploy and maintain an intervention over time.

The development of HAPIT was funded by the Global Alliance for Clean Cookstoves and was spearheaded by authors Ajay Pillarisetti and Kirk R. Smith of the Household Energy, Climate, and Health Research Group at the University of California, Berkeley. Early versions of HAPIT were created by Cooper Hanning. HAPIT benefits from the input of many colleagues, including Heather Adair-Rohani, Donee Alexander, Sophie Bonjour, Zoe Chafe, Manish Desai, Ellen Eisen, Maria Theresa Hernandez, L. Drew Hill, Nicholas L. Lam, and Lisa Thompson.

## References

Apte JS, Marshall JD, Cohen AJ, Brauer M (2015) Addressing global mortality from ambient PM2.5. Environ Sci Technol 49(13):8057–8066. doi:10.1021/acs.est.5b01236

Balakrishnan K, Ghosh S, Ganguli B, Sambandam S, Bruce N, Barnes DF, Smith KR (2013) State and national household concentrations of PM2.5 from solid cookfuel use: results from measurements and modeling in India for estimation of the global burden of disease. Environ Health 12(1):77. doi:10.1186/1476-069X-12-77

Bonjour S, Adair-Rohani H, Wolf J, Bruce NG, Mehta S, Pruss-Ustun A, Lahiff M, Rehfuess EA, Mishra V, Smith KR (2013) Solid fuel use for household cooking: country and regional estimates for 1980–2010. Environ Health Perspect 121(7):784–790. doi:10.1289/ehp.1205987

Bruce N, Weber M, Arana B, Diaz A, Jenny A, Thompson L, McCracken J, Dherani M, Juarez D, Ordonez S, Klein R, Smith KR (2007) Pneumonia case-finding in the RESPIRE Guatemala indoor air pollution trial: standardizing methods for resource-poor settings. Bull World Health Organ 85(7):535–544

Burnett RT, Pope CA 3rd, Ezzati M, Olives C, Lim SS, Mehta S, Shin HH, Singh G, Hubbell B, Brauer M, Anderson HR, Smith KR, Balmes JR, Bruce NG, Kan H, Laden F, Pruss-Ustun A, Turner MC, Gapstur SM, Diver WR, Cohen A (2014) An integrated risk function for estimating the global burden of disease attributable to ambient fine particulate matter exposure. Environ Health Perspect 122(4):397–403. doi:10.1289/ehp.1307049

Byass P (2010) The imperfect world of global health estimates. PLoS Med 7(11):e1001006. doi:10.1371/journal.pmed.1001006

Byass P, de Courten M, Graham WJ, Laflamme L, McCaw-Binns A, Sankoh OA, Tollman SM, Zaba B (2013) Reflections on the global burden of disease 2010 estimates. PLoS Med 10(7):e1001477. doi:10.1371/journal.pmed.1001477

Cairncross S, Feachem RG (1993) Environmental health engineering in the tropics: an introductory text, 2nd edn. Wiley, Chichester

Chang W, Cheng J, Allair J, Xie Y, McPherson J (2015) Web application framework for R. R RStudio, Boston, USA

GACC Country Profiles (2015) Global alliance for clean cookstoves. http://cleancookstoves.org/country-profiles/

EPA (2004) Advisory on plans for health effects analysis in the analytical plan for EPA's second prospective analysis – benefits and costs of the clean air act, 1990–2020. Scientific Advisor Board, United States Environmental Protection Agency, Washington, DC

EPA US BenMap: Environmental benefits mapping and analysis program user's manual. appendix. Office of Air Quality Planning and Standards, Research Triangle Park

Evans DB, Chisholm T, Edejer T (2006) Generalized cost-effectiveness analysis: principles and practice. Elgar Companion Health Econ. doi:10.4337/9781845428914.00061

Hill L, Pillarisetti A, Delapena S, Garland C, Jagoe K, Koetting P, Pelletreau A, Boatman M, Pennise D, Smith K (2015) Air pollution and impact analysis of a pilot stove intervention: Report to the Ministry of Health and Inter-Ministerial Clean Stove Initiative of the Lao People's Democratic Republic, Berkeley

IHME (2010) Global burden of disease study, ambient air pollution risk model 1990–2010. Institute for Health Metrics and Evaluation (IHME), Seattle

Jetter J, Kariher P (2009) Solid-fuel household cook stoves: characterization of performance and emissions. Biomass Bioenergy 33:294–305

Jetter J, Zhao Y, Smith KR, Khan B, Yelverton T, Decarlo P, Hays MD (2012) Pollutant emissions and energy efficiency under controlled conditions for household biomass cookstoves and implications for metrics useful in setting international test standards. Environ Sci Technol 46(19):10827–10834. doi:10.1021/es301693f

Johnson MA, Chiang RA (2015) Quantitative guidance for stove usage and performance to achieve health and environmental targets. Environ Health Perspect 123(8):820–826. doi:10.1289/ehp.1408602

Lozano R, Naghavi M, Foreman K, Lim S, Shibuya K, Aboyans V, Abraham J, Adair T, Aggarwal R, Ahn SY, Alvarado M, Anderson HR, Anderson LM, Andrews KG, Atkinson C, Baddour LM, Barker-Collo S, Bartels DH, Bell ML, Benjamin EJ, Bennett D, Bhalla K, Bikbov B, Bin Abdulhak A, Birbeck G, Blyth F, Bolliger I, Boufous S, Bucello C, Burch M, Burney P, Carapetis J, Chen H, Chou D, Chugh SS, Coffeng LE, Colan SD, Colquhoun S, Colson KE, Condon J, Connor MD, Cooper LT, Corriere M, Cortinovis M, de Vaccaro KC, Couser W, Cowie BC, Criqui MH, Cross M, Dabhadkar KC, Dahodwala N, De Leo D, Degenhardt L, Delossantos A, Denenberg J, Des Jarlais DC, Dharmaratne SD, Dorsey ER, Driscoll T, Duber H, Ebel B, Erwin PJ, Espindola P, Ezzati M, Feigin V, Flaxman AD, Forouzanfar MH, Fowkes FG, Franklin R, Fransen M, Freeman MK, Gabriel SE, Gakidou E, Gaspari F, Gillum RF, Gonzalez-Medina D, Halasa YA, Haring D, Harrison JE, Havmoeller R, Hay RJ, Hoen B, Hotez PJ, Hoy D, Jacobsen KH, James SL, Jasrasaria R, Jayaraman S, Johns N, Karthikeyan G, Kassebaum N, Keren A, Khoo JP, Knowlton LM, Kobusingye O, Koranteng A, Krishnamurthi R, Lipnick M, Lipshultz SE, Ohno SL, Mabweijano J, MacIntyre MF, Mallinger L, March L, Marks GB, Marks R, Matsumori A, Matzopoulos R, Mayosi BM, McAnulty JH, McDermott MM, McGrath J, Mensah GA, Merriman TR, Michaud C, Miller M, Miller TR, Mock C, Mocumbi AO, Mokdad AA, Moran A, Mulholland K, Nair MN, Naldi L, Narayan KM, Nasseri K, Norman P, O'Donnell M, Omer SB, Ortblad K, Osborne R, Ozgediz D, Pahari B, Pandian JD, Rivero AP, Padilla RP, Perez-Ruiz F, Perico N, Phillips D, Pierce K, Pope CA 3rd, Porrini E, Pourmalek F, Raju M, Ranganathan D, Rehm JT, Rein DB, Remuzzi G, Rivara FP, Roberts T, De Leon FR, Rosenfeld LC, Rushton L, Sacco RL, Salomon JA, Sampson U, Sanman E,

Schwebel DC, Segui-Gomez M, Shepard DS, Singh D, Singleton J, Sliwa K, Smith E, Steer A, Taylor JA, Thomas B, Tleyjeh IM, Towbin JA, Truelsen T, Undurraga EA, Venketasubramanian N, Vijayakumar L, Vos T, Wagner GR, Wang M, Wang W, Watt K, Weinstock MA, Weintraub R, Wilkinson JD, Woolf AD, Wulf S, Yeh PH, Yip P, Zabetian A, Zheng ZJ, Lopez AD, Murray CJ, AlMazroa MA, Memish ZA (2012) Global and regional mortality from 235 causes of death for 20 age groups in 1990 and 2010: a systematic analysis for the Global Burden of Disease Study 2010. Lancet 380(9859):2095–2128. doi:10.1016/S0140-6736(12)61728-0

Mathers CD, Lopez AD, Murray CJL (2006) The burden of disease and mortality by condition: data, methods, and results for 2001. In: Lopez AD, Mathers CD, Ezzati M, Jamison DT, Murray CJL (eds) Global burden of disease and risk factors. World Bank, Washington, DC

McCracken JP, Smith KR, Diaz A, Mittleman MA, Schwartz J (2007) Chimney stove intervention to reduce long-term wood smoke exposure lowers blood pressure among Guatemalan women. Environ Health Perspect 115(7):996–1001. doi:10.1289/ehp.9888

McCracken J, Smith KR, Stone P, Diaz A, Arana B, Schwartz J (2011) Intervention to lower household wood smoke exposure in Guatemala reduces ST-segment depression on electrocardiograms. Environ Health Perspect 119(11):1562–1568. doi:10.1289/ehp.1002834

Murray CJ, Ezzati M, Flaxman AD, Lim S, Lozano R, Michaud C, Naghavi M, Salomon JA, Shibuya K, Vos T, Wikler D, Lopez AD (2012) GBD 2010: design, definitions, and metrics. Lancet 380(9859):2063–2066. doi:10.1016/S0140-6736(12)61899-6

Northcross A, Chowdhury Z, McCracken J, Canuz E, Smith KR (2010) Estimating personal PM2.5 exposures using CO measurements in Guatemalan households cooking with wood fuel. J Environ Monit 12(4):873–878. doi:10.1039/b916068j

Pillarisetti A, Vaswani M, Jack D, Balakrishnan K, Bates MN, Arora NK, Smith KR (2014) Patterns of stove usage after introduction of an advanced cookstove: the long-term application of household sensors. Environ Sci Technol 48(24):14525–14533. doi:10.1021/es504624c

R Foundation: A language and environment for statistical computer. R Foundation for Statistical Computing, Vienna

Rosa G, Majorin F, Boisson S, Barstow C, Johnson M, Kirby M, Ngabo F, Thomas E, Clasen T (2014) Assessing the impact of water filters and improved cook stoves on drinking water quality and household air pollution: a randomised controlled trial in Rwanda. PLoS One 9(3):e91011. doi:10.1371/journal.pone.0091011

Ruiz-Mercado I, Masera O (2015) Patterns of stove use in the context of fuel–device stacking: rationale and implications. Ecohealth 12(1):42–56. doi:10.1007/s10393-015-1009-4. Epub 2015 Feb 28

Ruiz-Mercado I, Canuz E, Walker JL, Smith KR (2013) Quantitative metrics of stove adoption using Stove Use Monitors (SUMs). Biomass Bioenergy 57:136–148

Sambandam S, Balakrishnan K, Ghosh S, Sadasivam A, Madhav S, Ramasamy R, Samanta M, Mukhopadhyay K, Rehman H, Ramanathan V (2015) Can currently available advanced combustion biomass cook-stoves provide health relevant exposure reductions? Results from initial assessment of select commercial models in India. Ecohealth 12(1):25–41. doi:10.1007/s10393-014-0976-1

Singh A, Tuladhar K, Bajracharya A, Pillarisetti A (2012) Assessment of effectiveness of improved cook stoves in reducing indoor air pollution and improving health in Nepal. Energy Sustain Dev 16:406–414

Smith KR, Haigler E (2008) Co-benefits of climate mitigation and health protection in energy systems: scoping methods. Annu Rev Public Health 29:11–25. doi:10.1146/annurev.publhealth.29.020907.090759

Smith KR, McCracken JP, Thompson L, Edwards R, Shields KN, Canuz E, Bruce N (2010) Personal child and mother carbon monoxide exposures and kitchen levels: methods and results from a randomized trial of woodfired chimney cookstoves in Guatemala (RESPIRE). J Expo Sci Environ Epidemiol 20(5):406–416. doi:10.1038/jes.2009.30

Smith KR, McCracken JP, Weber MW, Hubbard A, Jenny A, Thompson LM, Balmes J, Diaz A, Arana B, Bruce N (2011) Effect of reduction in household air pollution on childhood pneumonia in Guatemala (RESPIRE): a randomised controlled trial. Lancet 378(9804):1717–1726. doi:10.1016/S0140-6736(11)60921-5

Smith KR, Bruce N, Balakrishnan K, Adair-Rohani H, Balmes J, Chafe Z, Dherani M, Hosgood HD, Mehta S, Pope D, Rehfuess E, Group HCRE (2014) Millions dead: how do we know and what does it mean? Methods used in the comparative risk assessment of household air pollution. Annu Rev Public Health 35:185–206. doi:10.1146/annurev-publhealth-032013-182356

Smith-Sivertsen T, Diaz E, Pope D, Lie RT, Diaz A, McCracken J, Bakke P, Arana B, Smith KR, Bruce N (2009) Effect of reducing indoor air pollution on women's respiratory symptoms and lung function: the RESPIRE Randomized Trial, Guatemala. Am J Epidemiol 170(2):211–220. doi:10.1093/aje/kwp100

UNDESA (2014) Revision of world urbanization prospects. United Nations Department of Economic and Social Affairs. http://esa.un.org/unpd/wup/

USCB (2015) International programs. United States Census Bureau. http://www.census.gov/population/international/

Van Vliet ED, Asante K, Jack DW, Kinney PL, Whyatt RM, Chillrud SN, Abokyi L, Zandoh C, Owusu-Agyei S (2013) Personal exposures to fine particulate matter and black carbon in households cooking with biomass fuels in rural Ghana. Environ Res 127:40–48. doi:10.1016/j.envres.2013.08.009

Voigt K, King NB (2014) Disability weights in the global burden of disease 2010 study: two steps forward, one step back? Bull World Health Organ 92(3):226–228. doi:10.2471/BLT.13.126227

WHO (2004) Indoor smoke from solid fuels: assessing the environmental burden of disease at national and local levels. World Health Organization, Geneva

Wickham H (2009) ggplot2. Springer, New York

Winfrey W, McKinnon R, Stover J (2011) Methods used in the lives saved tool (LiST). BMC Public Health 11(Suppl 3):S32. doi:10.1186/1471-2458-11-S3-S32

# Chapter 11
# Innovations in Payments for Health Benefits of Improved Cookstoves

**Ken Newcombe, Tara Ramanathan, Nithya Ramanathan, and Erin Ross**

**Abstract** This chapter presents two brief overviews of entrepreneurs demonstrating recent advances in monitoring and marketing the health impacts of improved cookstoves in developing countries. CQuest Capital proposes commoditizing and marketing Averted Disability Adjusted Life Years (ADALYs) associated with demonstrated reduced particulate exposure. Nexleaf Analytics leverages sensor technology and carbon credit markets to pay cookstove users for demonstrated use of stoves. Both efforts are innovations in the cookstove sector, and may over time contribute to improved cookstove program models.

**Keywords** Health credits • Carbon credits • Cookstoves • ADALY

## 11.1    From Carbon Credits to Health Credits

The arrival of a workable set of rules for Programs of Activities (PoAs), under the Clean Development Mechanism (CDM) in 2009, and the approval in 2008 of the small-scale project methodology AMS IIG to enable carbon crediting under the CDM for efficient cookstoves, made it possible to generate and trade typically 2–3 tonnes of $CO_2e$ avoided per efficient cookstove per year in most developing countries. In a healthy compliance market, with CDM's Certified Emissions Reductions (CERs) trading at Euro 10–15 per tonne of averted carbon emissions, it was profitable to distribute stoves at a marked discount to their delivered cost, cover the rigorous and time consuming requirements for monitoring and verification required by the methodology, and still make a profit. In this economic environment, there was no need to do more than mention in passing the cost-effectiveness of these interventions from a public health standpoint. These and other co-benefits of clean efficient

K. Newcombe (✉)
C-Quest Capital LLC, 1015 18th Street, NW Suite #730, Washington, DC 20036, USA
e-mail: knewcombe@cquestcapital.com

T. Ramanathan • N. Ramanathan • E. Ross
Nexleaf Analytics, 2356 Pelham Ave, Los Angeles, CA 90064, USA
e-mail: tara@nexleaf.org; nithya@nexleaf.org; erin@nexleaf.org

© Springer International Publishing Switzerland 2016
E.A. Thomas (ed.), *Broken Pumps and Promises*,
DOI 10.1007/978-3-319-28643-3_11

cookstoves compared to cooking over an open fire were simply embedded in project outcomes without measurement, or fanfare, along with the delivery of the climate benefit as verified through the issuance of CERs. In addition, the health benefits typically do not scale directly with fuel savings and would have to be evaluated separately.

The existence of non-monetized high economic value co-benefits of health and productivity improvements provides an opportunity to maintain investment in efficient cook-stoves. Even when carbon was $10/tCO$_2$e or more there was evidence that the economic value of the health benefits could outweigh the market value of the climate benefit. So one way to improve the financial viability and level of investment in improved cooks stove projects is to verify and monetize the health benefits for women and children. To do this the measurement of these health benefits must be simplified and standardized to establish the market mechanisms to sell them forward under a pay-for-performance (unit contingent payment) basis, much like the standardized practice in the carbon market of forward selling CERs from projects under an Emission Reductions Purchase Agreement (ERPA), including from improved cookstove projects. A plausible source of demand for this new health benefits commodity are nations, philanthropic foundations with a mandate to address these health issues, and high net worth individuals and families focusing their giving on women's and children's health in developing countries.

In this way, we can offer for forward purchase by investors the avoided health costs of inhalation of smoke from cooking on open biomass fire. The metric of improved health performance that would be sold forward is an Averted Disability Adjusted Life Year (ADALY), a standardized measure of population wide health performance from reductions in the burden of disease ranging from infectious disease to degenerative disease of environment origin (See Chap. 10). The contractual instrument would be a DARPA – a DALY Reductions Purchase Agreement. The DALY metric has the advantage of being widely understood and widely utilized as a measure of the cost-effectiveness of public health policies to address major causes of death and disease. The metric embodies both the qualitative and quantitative impacts of a disease or environmental health insult and is universal, addressing equally the poorest of the poor in the developing countries or wealthy individuals in industrialized countries.

Given that the economic benefits of reducing DALYs may potentially be higher than the climate benefits valued by the price of carbon, even modest improvements in health could have a significant impact on the economic viability of ICS projects if the health benefits can be monetized through market-based mechanisms.

## 11.2   Health Credits as Results Based Financing

Dedicated funding for improvement of health and well being in developing countries grew significantly in the early 2000s with establishment of the Global Fund to address Malaria, HIV Aids and Tuberculosis, the Global Alliance for Vaccination

(GAVI) with $300 million disbursed, and the World Bank's Results Based Financing (RBF) Health fund of $537 million committed through 2014. Growing attention to women's and children's health, especially in Africa, within RBF compared favorably to a voluntary carbon market annual trade of $400–$500 million only a small fraction of which supported clean cookstoves. All RBF requires independent verification agents (IVAs) and preferably peer-reviewed monitoring and verification protocols that assure financiers they are getting what they are paying for (see Chap. 2).

Developing a marketable product of Averted DALYs fits well with the direction of development assistance towards results based financing, or, as its sometimes called Pay-for Performance (P4P). Indeed, RBF was first practiced in the health sector associated with payments for the use of vaccines against infectious diseases, with the number vaccinated being the metric on which compensation was based. In development assistance for energy access, for example, including the use of improved cookstoves, the Norwegian Government, through NORAD has adopted RBF as a preferred standard.

The benefits of RBF include cost-effectiveness in the delivery of the benefits, enabling competitive allocation to identify the most cost-effective providers of the services and only paying on delivery for the desired outputs and outcomes: in this case Averted DALYs, or ADALYs.

A criterion for RBF is the ability to monitor and verify results against which payments are made. For the proposed product, indices need to be developed linking independently verified carbon emissions reductions issued into registries as a result of the use of specific cooking technologies (ICS) with Averted DALYs.

## 11.3   Generation and Verification

Some cookstove carbon project developers understood the rich "co-benefits" of cleaner more efficient cookstoves for the rural poor. Besides improved health outcomes from reduced smoke exposure, benefits include time-savings in fuel gathering and food preparation, productivity improvements, personal risk reduction from gathering cooking fuels, reduced pressure on valuable tree cover in surrounding landscapes, reduced caloric stress in times of food scarcity, and black carbon reductions as part of a range of additional climate mitigation and adaption benefits.

Two recent parallel developments support add materially to a clean stove-based health outcomes market by contributing to product integrity. These are:

*An effective CDM Efficient Stoves Methodology (AMS IIG):* the methodology, requires every stove to be recorded by serial number and location and a statistically rigorous survey to confirm that stoves exist and are in use. This is a key requirement for verifying the existence of health benefits arising from a clean cookstove intervention hence this part of AMS IIG can be readily adapted to an ADALY methodology;

*Launch of an on-line tool for estimating aDALYs and avoided premature deaths resulting from reduced kitchen air pollution:* The HAPIT tool, described in Chap. 10, can allow auditors to check the validity of claimed ADALYs and to verify their existence based on a statistically valid sample of measured personal PM 2.5 exposure levels resulting from cleaner cooking interventions. Clinical assessment of health performance in the target populations would be very expensive, logistically challenging, time consuming and wholly incompatible with market development at scale. HAPIT provides an affordable means of confirming and issuing for delivery and monetization ADALYs to health sector results based financiers under forward purchase contracts.

## 11.4   Economics

For illustrative purposes, the economics of a 50,000 stove project can be modeled based on our experience leading a stove intervention in Laos. Key assumptions include:

- Consumers would be prepared to pay $30/stove over 3–6 months;
- After-sales service system and parts over a 6 year proposed project life is $500,000 per year, the equivalent of the landed cost of the stove;
- Consumers would use the stove 50 % of the time.
- Kitchen air pollution is reduced fourfold compared to the traditional open wood-fired cooking arrangement;

On this basis the price of aDALYs needs to be about $2400, or about 1.4 times GDP/capita, to obtain a 20 % IRR unlevered. However, return expectations of private sector social impact investors for such a first-of-a-kind high-risk project is not known.

The World Health Organization (WHO) has established a guideline for cost-effectiveness that has been applied in the previously mentioned Laos research and is more broadly applicable to assessment of economic viability of clean cooking interventions. WHO recommends proceeding with interventions that reduce the burden of disease if they cost of aDALY generation is less than 3 times GDP per capita. Comparative analysis of opportunities to address a particular disease, the relative importance of that disease in terms of the overall burden of disease nationally, and opportunities to address the burden of disease in an economy generally bear heavily on whether to use scarce financial resources and institutional capacity on cleaner cooking.

Taking WHO guidelines into account, this Laos example suggests that ADALYs can be generated cost-effectively through use of forced draft "gasifier" cookstoves if they are adopted widely and used for more than half of all cooking. However, each country and cultural context will vary significantly in terms of adoption of various cookstove, ventilation and fuel intervention technologies, in the relative importance of biomass smoke exposure as a health problem, and relative cost-effectiveness of clean cooking interventions in addressing the national burden of disease.

## 11.5 Market

For forward sales of ADALYs to be the engine powering large-scale distribution of super clean stoves to the rural poor there must be creditworthy buyers on a matching scale. Who are the prospective buyers? There are at least three plausible categories: (i) OECD sovereigns through their bilateral aid agencies or through funds established in multi-lateral development finance institutions for output based aid that is either dedicated to women's and children's health improvements or where this objective is consistent with a broader mandate of results-based financing vehicles; (ii) private sector foundations using results-based financing with a focus on women's and children's health; and (iii) corporates with a commitment to social responsibility objectives that have chosen to support improved health outcomes for women and children, or that wish to contribute to these objectives as part of efforts to improve brand quality and supply chain stability. CQC is now collaborating with the World Bank and others to survey and engage potential ADALY buyers as part of ADALY market development.

## 11.6 Beyond Transparency: Enabling Results-Based Financing to Flow to Rural Women

Project Surya, led by University of California-San Diego, Nexleaf Analytics and TERI, promotes clean energy solutions that achieve health, climate, and sustainability benefits for the poorest three billion. Our goal is increasing adoption of clean cooking and lighting technologies that reduce household smoke emissions by 90 % or more. Effective distribution of clean cooking technologies as well as behavior change present serious challenges and are not well understood (Lewis and Pattanayak 2012); distribution, adoption, and ongoing maintenance are required to achieve emissions reductions. Financing these technologies presents another major challenge, as the more clean burning cookstoves cost more than many rural households can afford.

To that end, we use advanced energy technologies, proven in the field to reduce concentrations of black carbon and other harmful pollutants, to leverage the link between household pollution and climate change. By incorporating remote stove use monitoring into a household's improved cooking system, Surya provides a mechanism for users of advanced improved stoves to receive payments for their improved stove use. This approach is based in existing carbon market models that pay cash for reductions in climate change-inducing emissions. In our model, the individual women who cook on the improved stoves receive the carbon market-based financial benefits for their actions. This approach achieves a moral imperative, since the actors who reduce climate impact are the ones who receive the financing. It also dovetails with efforts to bolster stove user motivation for increased stove adoption, and it provides a fiscal mechanism for poor households to access and continue to maintain clean cooking technologies.

Carbon markets ensure a level of transparency and standardization of methods for verification and validation that will be important if this initiative is to scale up beyond Surya or any single institution. Surya is now working to expand this carbon market approach to encourage the adoption of clean lighting, as well as cooking, technologies. Close to a quarter of households now use the improved stoves for 50–100 % of their daily cooking needs. Each household that uses the stove for all cooking could earn approximately $35 per year (assuming an estimate of $6 per ton of $CO_2$ equivalent).

We leverage recent scientific advancements in understanding black carbon to create a new kind of carbon market aimed at encouraging and quantifying reductions in black carbon emissions. Black carbon particulates are one component of the particulate matter emitted by traditional solid biomass stoves. Black carbon in the air increases atmospheric warming, and is the second largest contributor to global warming after carbon dioxide (Ramanathan and Carmichael 2008). Black carbon particles carried to the Himalaya mountains come to rest on the snow and draw in more solar energy which accelerates melting of snow packs and glaciers (Menon et al. 2010). The snow melt combined with the atmospheric warming caused by black carbon contributes as much as 50 % to the anomalously large warming of the Tibetan-Himalayan glaciers (Bond 2013; Menon et al. 2010). About half of black carbon emissions from India are generated by cooking and other residential uses of biomass (Streets et al. 2013). Indeed, a single gram of black carbon warms the air at least as much as 350,000 g of $CO_2$ (Jacobson 2010), and has the equivalent climate effects of 3200 g of $CO_2$ evaluated over a 20-year time period (Bond 2013).

Given the deleterious impact of black carbon, reductions in black carbon, when included in carbon reduction calculations, can increase the number of tons of $CO_2$ and $CO_2$ equivalents valued on a carbon market. In 2015, the Gold Standard Foundation announced a new methodology, co-developed with Surya, that quantifies the emissions of black carbon and other short-lived climate pollutants produced when biomass is burned during cooking, and lays out methods for validating these reductions in these emissions which can leverage relatively low-cost technologies for BC monitoring in the field, thus providing a mechanism for arriving at a verified outcome. This mechanism for certifying reductions in black carbon, combined with the creation of a black carbon marketplace, will enable even more results-based financing to flow directly to rural women who increase their use of improved cookstoves.

# References

Bond TC (2013) Bounding the role of black carbon in the climate system: a scientific assessment. J Geophys Res 18(11):5380–5552
Jacobson MZ (2010) Short-term effects of controlling fossil-fuel soot, biofuel soot and gases, and methane on climate, Arctic ice, and air pollution health. J Geophys Res 115. doi:10.1029/200 9JD013795
Lewis JJ, Pattanayak SK (2012) Who adopts improved fuels and cookstoves? A systematic review. Environ Health Perspect 120(5):637–645. doi:10.1289/ehp.1104194

Menon S, Kock D, Beig G, Sahu S, Fasullo J, Orlikowski D (2010) Black carbon aerosols and the third polar ice cap. Chem Phys 10(10):4559–4571

Ramanathan V, Carmichael G (2008) Global and regional climate changes due to black carbon. Nat Geosci 1:221–227

Streets DG, Shindell DT, Lu Z, Faluvegi G (2013) Radiative forcing due to major aerosol emitting sectors in China and India. Geophys Res Lett 40:4409–4414

# Chapter 12
# The Role of Mobile in Delivering Sanitation Services

**Michael Ronan Nique and Helene Smertnik**

**Abstract** Lack of access to adequate sanitation has a considerable impact on public health, leading to mortality from diarrhea to stunting and malnutrition for children, directly affecting their school attendance. Open defecation, in addition to highly impacting privacy and dignity of individuals, is unsafe, especially for women, who are far more vulnerable to the risk of physical and sexual assault. Governments have the responsibility to improve sanitation capacity, but have often failed to provide service supply chain for operation and maintenance. Most practitioners also recognize that building a toilet cannot ensure that it will be used. In rural locations, sustainable access to sanitation is challenged by poor distribution networks, low availability to local sanitation solutions and open defecation behaviors. Growing mobile and data connectivity and use of mobile monitoring tools (SMS or applications), can improve understanding of community sanitation behaviors, while allowing service providers to develop more efficient supply chain and customer relationship management.

**Keywords** Sanitation • Mobile • Service

## 12.1 The Sanitation Gap

At least 2.5 billion people continue to lack access to improved sanitation globally, and over 70 % of these people live in rural areas of developing countries. Poor sanitation is associated with infectious disease, including diarrhea, helmith, and schistosomiasis. Diarrhea alone accounts for at least 1.4 million deaths annually, including nearly one out of five deaths in children under the age of five (Clasen et al. 2014). Given this severe health risk, significant efforts have been made particularly

M.R. Nique (✉) • H. Smertnik
GSM Association, GSMA, 2nd Floor, The Walbrook Building, 25 Walbrook,
London EC4N8AF, UK
e-mail: mnique@gsma.com

© Springer International Publishing Switzerland 2016                                       179
E.A. Thomas (ed.), *Broken Pumps and Promises*,
DOI 10.1007/978-3-319-28643-3_12

in the past 10 years to address poor sanitation coverage. In particular, India and Bangladesh have been the focus of extensive latrine coverage and utilization efforts in recent years.

Defined in 2000 by the United Nations and their member states, the Millennium Development Goals (MDG) 7 on "ensuring environmental sustainability" aimed to halve the proportion of people without access to "improved sanitation", defined as a facility that separates human excreta from human contact, by 2015 (WHO 2015). From 49 % of the world population with access to sanitation in 2000, the goal was to achieve 67 % of improved sanitation coverage by the end of this year. This target is not expected to be met, as the latest monitoring activities reported that 64 % of the world population had access to such infrastructure. Although two billion people gained access to improved facilities since 1990, 2.4 billion are still relying on "un-improved solutions", including flush toilets not connected to piped sewer system, septic tank or pit latrine; pit latrine without slab/open pit; bucket; hanging toilet or hanging latrine; shared facilities; no facilities (WHO 2015) and one billion practice open defecation. Moreover, as a result of population growth, there are more people without access to adequate sanitation today than in 2000 (Nothomb and Snel 2015).

Lack of access to adequate sanitation has a heavy impact on public health, leading to mortality from diarrhea to stunting and malnutrition for children, directly affecting their school attendance. Open defecation, in addition to highly impacting privacy and dignity of individuals, is unsafe, especially for women, who are far more vulnerable to the risk of physical and sexual assault (ETW 2014).

Closing the sanitation access gap is far from being an easy task and "lagging progress on sanitation persist because of the complexity of the response needed. Improving sanitation requires changing behavior and social norms (including open defecation)" (Wild et al. 2015b). Governments have the responsibility to improve sanitation capacity, but have often failed to provide service supply chain for operation and maintenance. Most practitioners also recognize that building a toilet cannot ensure that it will be used (Galvin 2015).

## 12.1.1  Progress on the Sanitation Millennium Development Goal

More than a third of the world's population, 2.4 billion people, has no access to "improved" sanitation solutions, including one billion people practicing open defecation (WHO 2015). Although important progress has been made globally, improving by 21 % since 1990 sanitation coverage in the developing world, the Millennium Development Goal (MDG) 7 on halving the number of people without access to sanitation by 2015 will not be met, with most countries of the Sub Saharan and South Asian regions clearly lagging behind (Fig. 12.1). Progress on sanitation has also often increased inequality by primarily benefitting wealthier people in urban areas, services declining sharply in informal settlements and rural environments (70 % of those without access to an improved sanitation facility live in rural areas).

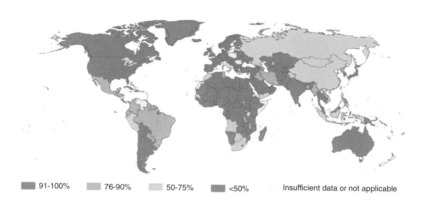

**Fig. 12.1** Proportion of the population using Improved Sanitation (UNICEF 2012)

Piped sewage systems and wastewater treatment plants only serve a fraction of those in developing countries, mainly due to the lack of political will, limited land availability and high cost of piped systems installation. As a result, a majority of the population (~2.6 Billion people (Smet 2014)) rely on on-site systems, such as pit latrines and septic tanks, requiring Faecal Sludge Management (FSM) services. The faecal sludge, a mix of solid and liquid waste, needs to be collected regularly to prevent the toilets from overflowing and contaminate its environment, such as water bodies. The lack of planning and commitment to maintenance of such systems are however leading to large portions of the waste being uncollected or untreated, with significant environmental, public health and economic impacts as mentioned above.

### 12.1.2   Transition to Sustainable Sanitation Services

Working within the new paradigm of the Sustainable Development Goals (SDGs) key elements that would support the development of sustainable sanitation services are (Nothomb and Snel 2015):

- More attention to the whole sanitation chain to move from a hardware delivery to a service delivery approach (Fig. 12.2) – sanitation is more than building a toilet and includes changed hygienic behaviours, maintenance, emptying, treatment and disposal or reuse of faecal matter;
- Clear leadership for change – sanitation improvements are not the sole responsibility of one entity, being usually spread between households, private service providers (latrine builders, emptying companies) and local and national governments;

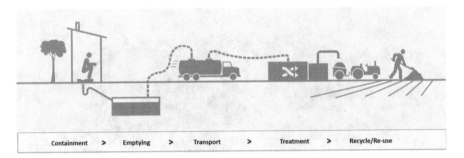

**Fig. 12.2** Sanitation value chain (Gates Foundation 2010)

- Unlocking public finance for implementing efforts at scale – the level of knowledge and understanding of financial flows to sanitation is very limited, due to the lack of reliable data tracking systems.

### 12.1.3 Improving Faecal Sludge Management in Urban Areas

With over 90 % of urban growth occurring in the developing world (World Bank 2011), an estimated 70 million residents are added to urban areas each year; this trend is especially important in South Asia and Sub-Saharan Africa, where the urban population is expected to double by 2030. As a result, slums are spontaneously emerging as a dominant and distinct type of settlement in fast growing cities, hosting 863 Million people (UN 2012) (or a third of the developing world's urban population). Although access to improved water and sanitation in slums has improved, with more than 227 million slum dwellers across Asia, Africa and Latin America gaining access to these basic services, more needs to be done.

The majority of urban dwellers do not have access to sewerage connections and rely instead on non-sewered onsite systems often shared with other families or public toilets. The capacity of such public solutions is however limited, as for example in Kibera, Nairobi largest slum, where 1000 public toilets are available for an estimated several hundreds of thousands people. Without access to facilities, people may defecate in the open or into a plastic bag (a "flying toilet").

In such low-income and slum environments, improving Faecal Sludge Management represents of the biggest challenges to achieving sanitation at scale. Without proper management, faecal sludge is often allowed to accumulate in poorly designed pits, is discharged into storm drains and open water, or is dumped into waterways, wasteland, and unsanitary dumping sites.

### 12.1.4 Increasing Access to Rural Sanitation Solutions and Changing Behaviours

While the Faecal Sludge Management problem is less critical than in urban areas, as rural dwellers have more flexibility to dig another pit to store their waste once a main pit is full, a whole sanitation chain approach would ensure waste is safely contained, but could also lead to the development of new business models, such as waste re-use for biomass or fertilizer, useful for local enterprises and communities.

Some of the main challenges to rural sanitation are access to local quality products and open defecation practices. Limited access to quality equipment from local entrepreneurs (e.g. masons able to build toilets) and their relative high cost compared to community's income are barriers to improved access. For local entrepreneurs, low population density, poor road access and low demand from rural communities, are also stifling the growth of such entrepreneurial sanitation enterprises.

## 12.2 Key Ecosystem Players

Both the public and private sectors, as well as local communities, are responsible for improving sanitation services in urban and rural areas, enabling these to grow through the implementation of sustainable business models.

**The Government: The Main Actor to Ensure Improved Sanitation Access**
In the sanitation sector, governments are often the main actors, responsible for the financing of sanitation services and creating demand. Their role is critical to create an enabling environment, in which on the one hand, policies exist to support the delivery of sanitation services including to the most remote areas and on the other hand, encourage the private sector's investments to bridge the sanitation gap (Perez 2012).

**Moving Towards Sanitation as a Business: The Role of the Private Sector**
With the support of engaged governments, entrepreneurs and service providers have the potential to transform the sanitation access sector, creating commercially viable and innovative business models and making the supply meet the growing demand for improved sanitation services. Universal access to improved sanitation services will require a concerted effort from both driven governments and the private sector, especially in light of the anticipated failure to reach the MDG sanitation target 2015.

**Communities and Non-Governmental Organizations: Providing Local Relevancy**
In order to ensure the good maintenance of sanitation facilities, suppliers (sanitation service providers, local government or non-government bodies) work on engaging local communities and giving them the responsibility to ensure the security and adequate maintenance of the facilities. Present on the ground, Non-Governmental Organizations (NGOs) can provide technical guidance, facilitate research, planning, design, capacity-building, implementation and monitoring.

## 12.3  Mobile Growth

While the MDG model of focalised international actions to increase living standards was failing to meet some of its targets, mobile phone ecosystems driven by entrepreneurial spirit and market forces managed to reach near ubiquity in many emerging markets: mobile networks now cover more than 85 % of the population and unique mobile subscribers' penetration is over 45 %. As a result, the gap between access to mobile and access to basic infrastructure such as utility services, has kept growing to the extent that, between 2002 and 2012, for every person gaining access to improved sanitation, ~2.5 persons became mobile subscribers.

More than mobile phone ownership, the level of sophistication of mobile services in many countries, such as mobile money, mobile internet and machine-to-machine connectivity, starkly contrasts with the status of sanitation services. For example in Kenya, where access to sanitation is reported at 30 % (UNICEF 2012), people are more likely to conduct financial transactions through their mobile money account (59 % of the adult population use mobile money) and browse the internet on their mobile phone (up to 40 % of the population), rather than benefit from the dignity, privacy and convenience of a well-maintained toilet.

This year, with the transition from the Millennium Development Goals to the Sustainable Development Goals aiming to set targets for the next 15 years, there is no doubt mobile devices, technologies and services have a role to play to support bridging the current infrastructure divide.

### 12.3.1  Using Mobile in the Sanitation Value Chain

As a result of the growing availability of mobile services and devices across urban and rural populations, mobile has become an increasingly interesting proposition in the sanitation sector. Although mobile integration in the sanitation value chain remains for now limited, its potential is high to bridge the existing divide between individuals, service providers and institutions, enabling data collection, remote monitoring, digitize information at the field level or provide innovative financing solutions.

### 12.3.2  Comparing Mobile and Sanitation Access

In terms of coverage, mobile access has outgrown access to improved sanitation in most emerging markets. The pace at which the number of mobile subscriptions grows is also much faster than the one of improved sanitation access: between 2002 and 2012, for every person gaining access to improved sanitation, ~2.5 persons

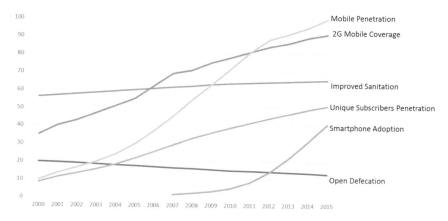

**Fig. 12.3** Mobile and sanitation indicators evolution (2000–2015)

became mobile subscribers (GSMA 2014a). As smartphones become increasingly available and affordable, their adoption is poised to disrupt mobile usage and the number of smartphones across the developing world is estimated to increase by 2.9 billion by out to 2020 (GSMA 2015).

Mobile coverage (through GSM networks) is now reaching more than 85 % of the emerging market population, making access to mobile services nearly ubiquitous in urban and rural settings with high population density. The remaining 15 % of the "un-covered" population is most likely to live in rural and remote areas, with low population density and sometimes difficult road access; in such locations, the business case for the deployment of mobile towers is often challenging for mobile operators and/or tower companies, due to high energy cost and low revenues. The ongoing technology migration to higher speed networks will also lead to more than 4 out 5 of people with access to 3G networks by 2020, up from 70 % today (GSMA 2015) (Fig. 12.3).

Mobile can play a role to support existing sanitation services, which reliability and sustainability could be improved, and upcoming services, to improve the sanitation coverage. Of the 3.1 Billion with people with access to sanitation services in the developing world, we estimate that 92 % are covered by mobile networks. In the context of un-improved sanitation and open defecation, e.g. 2.5 Billion people, we estimate that 72 % of this population, or up to 1.8 Billion people, respectively 691 Million and 1.1 Billion in urban and rural settings, are covered by mobile networks.

This divide between high mobile coverage and poor access to improved sanitation is especially strong in the regions of Southern Asia, South-Eastern Asia, Eastern Asia, Western and Eastern Africa. Overall, six countries account for more than 71 % of this total addressable market: India, China, Nigeria, Indonesia, Pakistan and Bangladesh (Fig. 12.4).

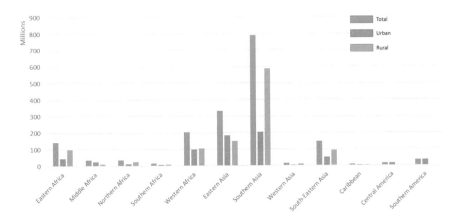

**Fig. 12.4** Sanitation addressable market (total, urban and rural) – number of people (millions)

### 12.3.3  A Fast Growing Smartphone Owners Base

Between 2000 and 2014, more than 5 Billion people gained access to mobile telephony around the world. If global mobile penetration growth is slowing down, market saturation has not been reached yet in many emerging markets, still building their subscribers' base and moving to value added and data services, fuelling the fast growth of global smartphone connections. As smartphones and mobile data plans become increasingly affordable, smartphone adoption is in a high growth phase in many markets, urban centers driving this growth; in 2014 alone, more than 612 million smartphone connections were added in developing regions.

As a result, in 2015, people might own a smartphone and browse internet on their mobile phone, but still don't have the dignity of a toilet and practice open defecation. In 33 countries, there are more unique subscribers than people using an improved sanitation solution, the majority of this population living in Sub Saharan Africa. In six countries (Kenya, Chad, Tanzania, Ghana, Nigeria & Namibia), there are more mobile internet connections than people with access to toilets.

### 12.3.4  Mobile Money Services Are Driving Financial Inclusion

Mobile money services are increasingly disrupting the financial landscape in emerging markets, allowing mobile subscribers to send remittances, pay their bills, save money or get access to micro-insurance plans through their mobile phone. There were more than 299 Million registered mobile money accounts, and more than 103 Million active accounts, in 2014 (GSMA 2014c) with 259 live mobile money

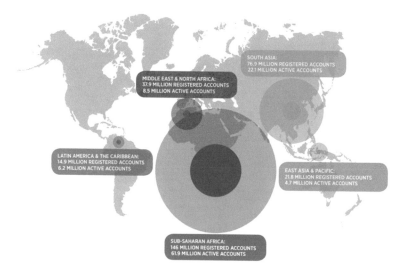

**Fig. 12.5** Registered mobile money accounts in 2014

services in 89 countries (Fig. 12.5). Beyond the East African countries which still host the largest mobile money deployments, 21 deployments worldwide have now more than 1 Million customers. Leapfrogging the formal financial infrastructure, 16 countries have now more mobile money accounts than bank accounts.

In East Africa, the number of registered mobile money accounts has reached more than 93 Million (GSMA 2014b), which is more than the number of people using improved toilets in this region. In Kenya where 30 % of the population has access to improved sanitation, according to current trends, it will take five generations to reach full sanitation coverage for the Kenyan population (Wild et al. 2015a). However, in less than a generation, mobile networks have been built to cover the majority of the country, smartphones have been increasingly distributed providing mobile internet access and more than 25 million mobile financial services subscribers are registered.

## 12.3.5  Mobile Integration in the Sanitation Value Chain

Focusing on the sanitation value chain stakeholders (households, entrepreneurs, NGOs & governments) have more opportunities to leverage mobile services, mobile money and M2M solutions to support the entire chain or specific elements from waste containment, emptying, transport, treatment and re-use (Fig. 12.6):

Below are examples in which communities, entrepreneurs, NGOs or governments, use mobile tools from toilet construction, to financing, and operations and maintenance.

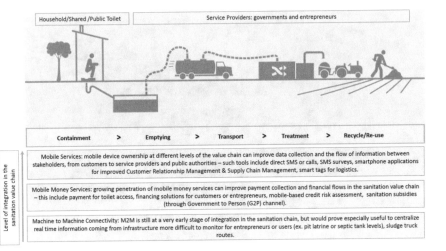

**Fig. 12.6** Mobile in the sanitation value chain (Gates Foundation 2010)

### 12.3.6 Improving Access to Local Sanitation Builders

Connecting supply and demand is an important challenge to improve access to quality sanitation infrastructure in rural environments. For communities in demand of a private or shared toilet, the lack of local sanitation entrepreneurs and poor distribution networks can lead to long delays before construction. IDE in Cambodia (iDE-Cambodia 2015) has been supporting private sector initiatives since 2008, developing access to quality toilets in local markets at an affordable price (below <US$50), leading to a fourfold sales increase since 2008 (Shah et al. 2014). IDE has also been using Salesforce as a Customer Relationship Management tool, with an Android-based application Taroworks (Taro 2015) for mobile data collection, to support local teams efficiency and interactions with customers. The team is also looking to adopt a more sophisticated mobile solutions which would be able to handle transactions, Supply Chain Management as well as the Monitoring & Evaluation functions.

### 12.3.7 Financing Sanitation Solutions

The upfront cost of sanitation infrastructure and equipment is a significant obstacle to improving access in urban and rural areas. To overcome this financial barrier, NGOs or social enterprises such as Sanitation Solutions Group in Uganda (SSG 2015) or Sanergy in Kenya (Sanergy 2015), offer payment plans for entrepreneurs to access financing solutions through local financial institutions partnerships. Financing access to sanitation infrastructure can be however viewed by financial institutions as income enhancer, not income generator, preventing such credit lines to scale and be available for most.

Recent studies show that repayment rates and impact could be high: in India in 2011, more than 146,000 toilet loans were provided to low income households, with high repayment rates (~98 %) (Trémolet 2013). Another recent research on sanitation microfinance in Tanzania nonetheless shows that "much remains to be done to scale up microfinance for sanitation, and more research is needed, particularly for assessing the impact of contracting a loan for sanitation on households' health and financial situation" (Mansour 2015). In markets where mobile money services are dynamic, mobile phones could offer efficient payment collection tools while enhancing payment plan flexibility.

New types of data such as mobile-phone usage patterns, airtime top-ups or mobile payments also offer new opportunities for financial institutions and lenders to have a more complete understanding of households' financial behaviours and needs, while developing new credit risk assessment models. Mobile operators, utility companies (this includes solar Pay As You Go providers) and data analytics companies (First Access, Lendable) are using new approaches to tap into the new forms of data spun off from mobile usage to build better risk models. Mobile payments data can provide credit underwriters with rich transactional information for generating credit insights. Having a better understanding of customers credit profiles would also lead to lower interest rate as customer is less likely to default.

Telenor, an international mobile operator group operating in 29 countries, is developing new types of mobile credit products by building predictive credit-scoring models in-house. These solutions are currently being piloted in Thailand for mobile money loans, emergency airtime top ups and handset financing. Once this credit risk assessment model is validated, it will be interesting to see if financial institutions can rely on this model to provide loans of higher value to mobile subscribers willing to get access to a utility infrastructure.

## 12.3.8   Improved Monitoring

Although investments have been flowing in the water and sanitation sector (financial commitments increased by 30 % between 2010 and 2012, from $8.3 billion to $10.9 billion (WHO/UNICEF 2014)), a category of people do not use the toilets built, due to cultural (ex. open defecation) or operational factors (ex. poor maintenance). In India, the government has started to deploy a nation-wide campaign to monitor sanitation activities using mobile tools in order to ensure a sustainable approach to sanitation access (Hueso 2014).

As part of such governments strategy to improve sanitation coverage and reduce open defecation, mobile-based monitoring activities have been piloted in several countries in recent years: for example, by using mobile applications on smartphones or tablets to verify that communities are maintaining their open defecation free status (World Bank 2014) in India or through SMS-monitoring system in Indonesia

(Wold Bank 2014). Machine to Machine connectivity has also been piloted at a smaller scale, based on the integration in rural community latrines of a GSM connected passive infrared motion detector, microcontroller, memory card and battery developed by SWEETLab of Portland State University. Similarly to man-made structured observations, the results showed that such M2M solutions proved a promising technology to provide detailed measures of latrine use for a better understanding of sanitation behaviors.

Beyond data collection, current pilots and small scale deployments have also outlined that there is a need for a harmonized approach to rural sanitation, clear monitoring framework to measure performance and the need for multiple data verification systems to ensure data accuracy. For governments and service providers, it's also about providing timely response and action to support operations and maintenance of sanitation infrastructure in order to keep the community involved and confident in such monitoring processes.

### 12.3.9 Digitizing Payments and Enhancing Productivity of Sanitation Entrepreneurs Digitalizing Payments

While domestic Peer to Peer (P2P) transfers and airtime top-ups continue to dominate the global mobile money product mix, the fastest growth in 2014 occurred in bulk disbursements, bill and merchant payments (GSMA 2014c). In slums, where storing and carrying cash can be risky, mobile money enable subscribers to safely store their money online, after a cash-in operation through their local mobile money agent. In growing mobile money markets, there is increased opportunity for sanitation solution providers to offer digital payment options to their customers, limiting cash handling and allowing service providers and customers to keep digital records of payments.

Some limits however exist on mobile money low value transactions, especially relevant for private or public toilet payments. In some mobile money deployments, the minimal amount a subscriber can pay or transfer is usually above $US0.10; the fees charged on each transactions (which contribute to mobile money agent revenues) can also be too high to make mobile payments interesting in this segment (toilet access usually costs below US$0.10). Because of these mobile money limitations, sanitation entrepreneurs such as the Umande Trust in Nairobi have been piloting cashless payments, allowing residents to pay for toilet access with pre-paid smart cards. Using Near Field Communication (NFC) enabled Android-based smartphones as Points-Of-Sale, customers are estimated to pay a 1 % fee per transaction, lower than competitors in the mobile payment space (Auerbach 2014). Based on the number of transactions processed by such sanitation entrepreneurs, an interesting proposition to pilot with Mobile Network Operators would be to offer low cost or free transactions on this very low tier (sub US$0.10) (Kumar and Toru 2011), to better understand how this payment category (low value high frequency) can contribute to mobile customer retention while supporting entrepreneur productivity.

### 12.3.10 Mobile Productivity Applications for Entrepreneurs

Data collection and geo-mapping tools (Kopernik 2015), customer relationship management (CRM) and supply chain management (SCM) applications further enable the digitization of an information collected at the field level. Open Data Kit (ODK) tools, free and open-source set of tools, can be used in combination with cloud based CRM solutions to push data directly in an organization database and help automatize process. More social enterprises are now using such enterprise applications to track their business and agents performance from marketing, waste collection, toilet usage or entrepreneur income generation.

### 12.3.11 Optimizing Emptying Services Call Centre in Dakar for Mechanized Desludging Services

In Senegal, the public sanitation service organization (ONAS) has developed, in partnership with local software company Manobi, a call center enabling urban households in Dakar to call in when in need of a desludging service (USAID 2014). This service aimed to improve linkages between service suppliers and customers, but also support the development of a sanitation private sector and enhance the demand for mechanical service. Once an emptying call is registered, desludging operators can submit an offer and calls for bids go out over SMS, this requiring a low level of technology. The lowest bidder amongst the operators win the job. The service has been live since March 2014, so far leading to an increase in the number of mechanical desludging services provided and a slight decrease in the price of emptying services.

### 12.3.12 Smart Sensors to Automate Emptying Services Request

Smart sensors are already in use in developed country cities to track the activity of sewerage networks in order to prevent blockages. In the case of lower income settings and on site systems, there is need to develop a more cost efficient solution for smart sensor integration in sanitation infrastructure. M2M connectivity, where sensors combined to a wireless connectivity chipset (GSM or other shorter range wireless technologies), could improve access to real time information on parts of the on-site sanitation infrastructure more difficult to monitor by humans: for example, pit and septic tank levels. Such information would prove useful in commercial sanitation services where automated messages could help entrepreneurs provide timely responses, also able to better organize their route based on providers and customers location.

### 12.3.13  Improving Sludge Logistics Management

As well as being poorly managed, waste transport or logistics is one of the biggest cost factors in the faecal sludge management chain. From waste collection to the treatment facility (when available), there is little incentive for trucks to dispose the sludge at the plant level; as a result, waste is often disposed into environment and often contaminate local water bodies. One of the mobile tools being currently piloted is about enabling objects and infrastructure to become "passively" smart: information can be embedded in smart RFID tags or QR codes which can be scanned by mobile devices, connecting an object/infrastructure/location to an online database or "object hyperlinking". Such solution can prove useful to accurately monitor toilet servicing and maintenance but also improve logistics for toilet distribution or emptying services. The Water and Sanitation Program of the World Bank in Indonesia has recently piloted the combined use of Android smartphones and QR codes to improve sludge trucks management. By scanning codes at the base, household and treatment plant locations, emptying services providers are improving information collection about their customers' base, while this process also aims to prevent illegal dumping of sludge and improve asset management.

### 12.3.14  Innovation Funding for Sanitation Service Providers

Data is critical to sanitation service providers and global organizations to monitor progress and a way for donors and investors to better understand the impact of the for-profit or non-profit organizations being financed. In that light, even though investments in the sanitation sector have increased in recent years, there is still a considerable gap between the demand and capital for funding in the sanitation sector. As many traditional approaches to WASH funding have not worked, local entrepreneurs and social enterprises are emerging as important players to improve capacity and efficiency of existing services. Such social enterprises can receive seed funding from organizations such as USAID DIV and Gates Foundation, which recently created the WASH for Life Partnership, to identify and test new WASH technologies and delivery models, scaling and replicating those that prove successful (Zeilberger 2015).

As presented in the report "How to Spend a Penny", outlining the Stone Family Foundation (SFF) innovation funding strategy to entrepreneurial initiatives in the WASH sector (Plimmer et al. 2014), WASH is not a sector for quick wins and access to patient capital is needed to allow models to be refined before scaling. It is more than building toilets and taps, and the changes to attitudes and behavior required to increase access to effective WASH services takes time. Access to innovative financing solutions will also improve households' ability and willingness to pay for WASH.

## 12.4  Next Steps for Mobile in the Sanitation Sector

Digital landscapes in emerging markets are very different now than it was 15 years ago, when the Millennium Development Goals were first articulated. In the transition to the Sustainable Development Goals setting development targets for the next 15 years, Information and Communication Technologies have a strong role to play in strengthening but also monitoring progress. While mobile is at an early stage of integration in the sanitation sector, mobile technologies and services could play an important role to support public and private initiatives at different stages of the value chain in both urban and rural settings.

In urban contexts, where population is growing fast, often localized in underserved zones and putting increasing pressure on municipalities to provide utility services, widespread mobile phone penetration and growing mobile money services uptake in some markets are giving the opportunity to utilities and sanitation solution providers to increase efficiency of services, while building new engagement models with customers.

In rural locations, sustainable access to sanitation is challenged by poor distribution networks, low availability to local sanitation solutions and open defecation behaviors. Growing mobile and data connectivity and use of mobile monitoring tools (SMS or applications), can improve understanding of community sanitation behaviors, while allowing service providers to develop more efficient supply chain and customer relationship management.

Although a clear willingness from organizations and enterprises to pilot mobile tools has been identified, challenges remain about the types of tools available and more importantly how to integrate them. Considering the potential of mobile, grant capital is first needed to encourage and support pilots and the development of tailored mobile solutions, such as mobile communication platforms, low cost M2M solution or mobile payments. Partnerships with technology providers (software and hardware) and mobile operators would also strengthen the development of such solutions, while impacting both businesses.

## References

Auerbach D (2014) World toilet day – discussing the potential power of mobile payment systems for the sanitation sector. GSMA, London

Clasen T, Boisson S, Routray P, Torondel B, Bell M, Cumming O, Ensink J, Freeman M, Jenkins M, Odagiri M, Ray S, Sinha A, Suar M, Schmidt WP (2014) Effectiveness of a rural sanitation programme on diarrhoea, soil-transmitted helminth infection, and child malnutrition in Odisha, India: a cluster-randomised trial. Lancet Glob Health 2(11):e645–e653. doi:10.1016/S2214-109X(14)70307-9

ETW (2014) Sanitation matters: why and how we are empowering rural women to build toilets. Embracing the world. Broken Pumps and Promises – Unlocking Pay for Performance in Environmental Services10.9.15.docx, India

Galvin M (2015) Talking shit: is Community-Led Total Sanitation a radical and revolutionary approach to sanitation? WIREs Water 2:9–20. doi:10.1002/wat2.1055

Gates Foundation (2010) Water, sanitation & hygiene. Bill & Melinda Gates Foundation, Seattle

GSMA (2014a) GSMA M4D utilities. GSMA, London

GSMA (2014b) GSMAi mobile money accounts in East Africa. GSMA, London

GSMA (2014c) State of the industry: mobile financial services for the unbanked. Mobile Money for the Unbanked. GSMA, London

GSMA (2015) The mobile economy. GSMA, London

Hueso A (2014) Will Narendra Modi free India from open defecation? Water Aid. http://www.wateraid.org/news/news/will-narendra-modi-free-india-from-open-defecation

iDECambodia (2015) iDECambodia. http://ide-cambodia.org/

Kopernik (2015) Digital data collection apps. Kopernik. http://impacttrackertech.kopernik.info/digital-data-collection-apps

Kumar K, Toru M (2011) Can mobile money be "Free"? CGAP, Washington, DC. Broken Pumps and Promises – Unlocking Pay for Performance in Environmental Services10.9.15.docx

Mansour G (2015) Lessons learned from the sanitation of microfinance action-research. Sanitation and Hygiene Applied Research for Equity. http://www.shareresearch.org/NewsAndEvents/Detail/Sanitation-Microfinance-Lessons-Learned

Nothomb C, Snel M (2015) Let's face the sanitation chain challenges together. International Rescue Committee (IRC). Broken Pumps and Promises – Unlocking Pay for Performance in Environmental Services10.9.15.docx

Perez E (2012) What does it take to scale up rural sanitation. World Bank, Washington, DC

Plimmer D, Baumgartner L, Hedley S (2014) How to spend a penny. NPC and The Stone Family Foundation, London

Sanergy (2015) Sanergy. http://saner.gy/

Shah NB, Shirrell S, Fraker A, Wang P, Wang E (2014) Understanding willingness to pay for sanitary latrines in Rural Cambodia: findings from four field experiments of iDE Cambodia's Sanitation Marketing Program. IDinsight, San Francisco

Smet J (2014) Faecal sludge management: the forgotten link in the sanitation service chain. IRC, The Hague

SSG (2015) Sanitation Solutions Group. https://sanitationsolutionsgroup.com/

Taro (2015) Taro works: making data work for you. http://taroworks.org/

Trémolet S (2013) Sanitation mircrofinance: a solution to the household sanitation cash trap? Sanitation and Hygiene Applied Research for Equity. https://sanitationsolutionsgroup.com/

UN (2012) Global urban indicators database. UN-Habitat, New York

UNICEF (2012) Progress on drinking water and sanitation. Joint monitoring program. Unicef and World Health Organization, New York

USAID (2014) Using technology to optimize Dakar's sludge market. USAID. http://usaid-suwasa.org/index.php/component/k2/item/310-using-technology-to-optimize-dakars-sludge-market

WHO (2015) Joint Monitoring Program (JMP) for water supply and sanitation. World Health Organization and UNICEF. http://www.wssinfo.org

WHO/UNICEF (2014) Progress on drinking water and sanitation: joint monitoring programme update 2014. WHO/UNICEF, Geneva

Wild L, Booth D, Cummings C, Foresti M, Wales J (2015a) Adapting development: improving services for the poor. UK Aid, London

Wild L, Booth D, Cummings C, Foresti M, Wales J (2015b) Adapting development: improving services for the poor. UK Aid, London

World Bank (2014) Scaling up Indonesia's rural sanitation mobile monitoring system nationally water and sanitation program. World Bank Group, Washington, DC

World Bank (2014) On world toilet day, focus is on equality and dignity. The World Bank. http://www.worldbank.org/en/news/feature/2014/11/18/on-world-toilet-day-focus-is-on-equality-and-dignity

World Bank (2011) World development report. World Bank, Washington, DC

Zeilberger C (2015) Thinking outside the latrine: startups test new toilets USAID. http://www.usaid.gov/global-waters/march-2015/thinking-outside-latrine?utm_source=Newsletter+-

# Chapter 13
# Combining Sensors and Ethnography to Evaluate Latrine Use in India

**Kathleen O'Reilly, Elizabeth Louis, Evan A. Thomas, and Antara Sinha**

**Abstract**  This chapter presents recent research in latrine use measurements—a challenging element of sanitation service delivery. The research used quantitative and qualitative methods to contribute to new understanding of sanitation practices and meanings in rural India. We estimated latrine usage behavior through ethnographic interviews and sensor monitoring, specifically the latest generation of infrared toilet sensors. Two hundred and fifty-eight rural households in West Bengal (WB) and Himachal Pradesh, India, participated in the study by allowing PLUMs to be installed in their houses for a minimum of 6 days. Six hundred interviews were taken in these households, and in others, where sensors had not been installed. Ethnographic and observational methods were used to capture the different defecation habits and their meanings in the two study sites. Those data framed the analysis of the PLUM raw data for each location. PLUMs provided reliable, quantitative verification. Interviews elicited unique information and proved essential to understanding and maximizing the PLUM data set. The combined methodological approach produced key findings that latrines in rural WB were used only for defecation, and that low cost, pit latrines were being used sustainably in both study areas.

**Keywords**  Behavior change • Ethnography • India • Policy • Sanitation • Sensor monitoring

Portions adapted and reprinted from Journal of Water Sanitation and Hygiene for Development (2015) 5(3), 426–438, doi: 10.2166/washdev.2015.155 with permission from the copyright holders, IWA Publishing.

K. O'Reilly • E. Louis
Department of Geography, Texas A&M University, 810 Eller Building, MS-3147,
College Station, TX 77843-3147, USA

E.A. Thomas (✉)
Department of Mechanical and Materials Engineering, Portland State University,
1930 SW 4th Ave, Portland, OR 97201, USA
e-mail: evthomas@pdx.edu

A. Sinha
Faculty of Infectious and Tropical Diseases, London School of Hygiene
and Tropical Medicine, London, UK

© Springer International Publishing Switzerland 2016
E.A. Thomas (ed.), *Broken Pumps and Promises*,
DOI 10.1007/978-3-319-28643-3_13

## 13.1  Introduction

Increased latrine coverage has generally been the primary metric used to evaluate the impact of sanitation interventions in Bangladesh, India and elsewhere. In this regard, many programs have been successful. In one recent study, the intervention increased latrine coverage from 9 to 63 %, compared to a control group that increased from 8 to 12 %. However, the intended health impact was not subsequently realized. The prevalence of diarrhea in the intervention was 8.8 %, while the control group was 9.1 %, and mortality estimates were roughly similar as well (Clasen et al. 2014). This study suggested that latrine coverage was an insufficient metric, and that utilization of latrines is a more appropriate measure that is more closely aligned with health impacts.

Measuring use has historically been challenging. Numerous studies have shown a respondent bias, and structured observations, previously the gold standard approach, have now been demonstrated to be highly reactive. Therefore, improved, objective utilization methods are required. For example, data from a recent study conducted in Bangladesh demonstrated an upward bias in the difference between respondent-reported 'likely defecation' events and sensor instrument-recorded events relative to the average between the measures. These findings indicate an over-estimation of respondent-reported latrine utilization relative to instrument-recorded use. The average difference between respondent-reported and instrument-recorded events indicated an average of 11 excess respondent-reported events (95 % CI 53, -30). The concordance correlation coefficient (CCC) between respondent-reported and instrument-recorded utilization was 29 (95 % BCa CI 0.15, 0.43). This CCC indicated that respondent-reported 'likely defecation' events were only weakly correlated with instrument-recorded 'likely defecation' events. While there was a moderately high level of accuracy in the measures, the data were imprecise, as indicated by the broad spread of observations from the reduced major axis (Delea et al. 2015).

This exaggerated self-reported use raises serious questions about the accuracy of self-reported data often used for policy and programmatic decision-making. Critically, the metrics used by program funders and implementers must at minimum narrow the gap between inputs and impact. While use may not be a sufficient measure, it is clear that measuring coverage alone is insufficient. Electronic sensors may improve the objectivity of latrine use measurement, and enable more continuous monitoring. Sanitation studies have yet to resolve the question of how to measure toilet usage with accuracy and sensitivity, leaving open the question of whether current policy is effective (Cousens et al. 1996; Rodgers et al. 2007). As Thomas et al. (2013) recommended, more rigorous, innovative evaluations are needed to guide best practices and improve future programs. Without clarity on why sanitation is adopted in some places and not others, programming and policy development is made more difficult.

This paper intends to fill a gap in studies of rural sanitation by demonstrating the combined strengths of quantitative and qualitative methods. We used Passive Latrine Use Monitors (PLUMs; instrumented monitoring) to quantify toilet usage. We used ethnography to learn about users, their beliefs about sanitation, and how beliefs influenced practices (Rheinlander et al. 2010). Ethnography is judged methodologically by different criteria than quantitative methods (Small 2009), leading to some tensions in research design. However, combining the two methods enabled insights into everyday sanitation behavior, including key findings that: (1) toilets across the WHO/UNICEF Joint Monitoring Program for Water Supply and Sanitation (JMP) spectrum were sustainably used in both study areas; and (2) beliefs of impurity limited toilet use to defecation in West Bengal. We discuss these findings below, after a brief review of the literature.

## 13.2    Understanding and Monitoring Sanitation Adoption

Studies deploying ethnographic methods, especially in-depth interviews, have uncovered a number of non-health related reasons motivating toilet building, e.g., social prestige, protection of women family members, desire to be modern, desire to take advantage of something given with little opportunity cost to the family, and rising household incomes (Jenkins and Curtis 2005; Jenkins 2004; Srinivas 2002; O'Reilly and Louis 2014). Interviews and focused group discussions have illuminated geographic variations in meanings of waste and hygiene; local norms for gendered, age-relevant defecation practices; and socio-religious rules about waste disposal matter for sanitation uptake (Drangert and Nawab 2011; McFarlane 2008; O'Reilly 2010). As Rheinlander et al. (2010) argued, knowledge of communities' beliefs about defecation is critical, as practices derive from beliefs. Insights into beliefs, values and meanings may be learned by asking people about them, and by observing their practices as a reflection of their beliefs. We used ethnography to illuminate geographically-specific toilet use behaviors and beliefs behind them.

Researchers have tackled the problem of assessing toilet usage (Olsen et al. 2001; Montgomery et al. 2010), but as yet, no single observational solution manages to be accurate, sensitive and non-intrusive. Structured observation at peak times of toilet usage is intrusive and may alter users' behavior (Clasen et al. 2012; Ram et al. 2010). It is also time-consuming, costly, and therefore difficult to scale up, while only providing a limited snapshot of potentially biased behavior. Observational methods such as looking for fresh feces in the pit or in open defecation areas, presence of materials for anal cleansing, and/or a wet toilet floor are subjective, lack sensitivity and specificity, and may be impossible given the toilet technology (Clasen et al. 2012). Self-reporting is also problematic as individuals may over-report in an effort to please the data collector, and gender of the evaluator has been shown to cause under-reporting (Manun'Ebo et al. 1997).

Cellular phone network-based monitoring technology has been field-tested to record usage and behavior change in WASH and other public health interventions, e.g., the provision of household water filters, hand washing stations, and cookstoves (Thomas et al. 2013). Effective use of remote monitoring is made possible by improved cellular networks, low cost of electronic components, and improved battery technology (Thomson et al. 2012; Thomas et al. 2013). The main argument for using electronically instrumented monitoring technologies is that they provide cost-effective, objective, accurate, regular, and continuous data thereby filling a critical gap in the ability to monitor health interventions effectively (Thomas et al. 2013; Clasen et al. 2012).

Below we discuss the study site and population selection rationale before moving into the specific methods guiding the quantitative and qualitative portions of the research. An analytical section follows, including a description of our iterative process, and discussion of findings. We conclude that, despite the challenges of integrating disparate methodological tools, combined methods offer new understandings of sanitation behavior in rural India.

## 13.3    Site Selection and Study Population

Our goal was to contribute new insights into effective sanitation by studying unique places where sanitation was adopted at rates of almost 100 % in parts of rural India. Therefore, the research was conducted in rural villages areas of West Bengal (WB) and Himachal Pradesh (HP)—two geographically and economically different states that have made some of the greatest improvements in sanitation coverage in the past 20 years (Table 13.1).

We chose Gram Panchayats (GPs; i.e., political subdivisions comprising multiple small villages) that won the Clean Village Award (NGP; a cash award for open defecation free status) in the past 3–5 years and that were well-known locally and extra-locally as areas of high toilet usage. Selected GPs were of mixed caste and class composition to enable a broad, socio-demographic cross-section of participants. Several individual household latrine (IHL) types were observed at each site; most were improved sanitation. Toilet cabins ranged from plastic sheeting to brick and mortar walls with slab roofs. Almost all toilets were built at a distance from the main dwelling. In HP, some households had attached (to the house) toilets in a room large enough for bathing (hereafter, toilet/bathroom).

## 13.4    Quantitative Methods – Sensor monitoring

The technology employed in this study, Portland State University Passive Latrine Use Monitors (PLUMs), is described in technical detail in other publications, including Thomas et al. 2013. A simple infrared motion detector was used, identical

**Table 13.1** Percentage of households without toilets in WB and HP – 1992/1993–2011

| State | 1992/1993 | 2001 | 2011 |
|-------|-----------|------|------|
| WB | 59.6 | 56.3 | 41.2 |
| HP | 87.4 | 66.6 | 30.9 |
| All India | | | 54.3 |

to the commercial sensor selected in the Clasen et al. (2012) study. A comparator circuit was linked with the motion detector, and recorded each detected motion. One or more times per day, the comparator board relayed logged data events to the internet via GSM cellular technology. A handheld cell phone was used to determine if a signal could be located at the household, indicating the PLUM could communicate with the cell phone tower. If a strong signal was unavailable, it was switched into "local" logging mode on a micro-SD card and data was manually uploaded after removal from the toilet. PLUMs were fastened with zip ties (aka cable ties) within 5 ft of the toilet pan.

Forty PLUMs were utilized and were rotated between 291 households. In related studies, PLUMs suggested low behavioral reactivity after the first several days, so PLUMs were installed for 7–10 days to capture behavior for at least 6 days of data. PLUM installations occurred based on willingness to accept, and the presence of the household head. The PLUM installation sample illustrates one of the tensions arising from combining qualitative and quantitative methods: we do not claim a representative, random, or unbiased sample of households with PLUMs installed. Ethical obligations prevented the installation of PLUMs in households that refused them, which may have biased the data if refusal was due to toilet non-use. However, respondents were forthcoming in interviews about household members who went for open defecation whether they accepted PLUMs or not, nor was there a noticeable difference in PLUM acceptance across the study sites once we routinized our installation strategy. Informants' honesty also enabled us to better calculate the number of toilet users per household, refining PLUM data analysis. It is possible that interviewing before installation and the initial presence of the PLUM may have influenced household behavior. This potential reactivity has not been rigorously characterized to date.

The PLUM online software system contains several data correction, reduction and analysis routines. Subsequently, an R code is run to interpret the raw data and generate estimates of 'usage events'. The algorithm employed is largely based on Clasen et al. 2012, with some adjustments to account for technological differences between the sensors.

## 13.5 Qualitative Methods – Ethnography

We conducted over 600 in-depth semi-structured interviews with household members and key informants. The rationale for 600 interviews was to insure saturation (i.e., interviews produced no new data) and to interview across socio-economic characteristics and toilet type in each of the four GPs. We only interviewed in households where toilets were present and householders reported that they were being used. Respondents were adults, but not necessarily the household head. Household interviews covered: family composition, general usage, household toilet building history, and their understandings of human waste, sanitation, and hygiene. We did not ask respondents about their usage habits because we found early in the field period that respondents grew suspicious that we were 'checking' (i.e., official record keeping that may have negative repercussions for households) on toilet usage. Households were reassured that we were not 'checking,' but seeking to confirm our information that these were GPs where most households used their toilets. This strategy of reassuring interviewees highlights again the tensions between qualitative and quantitative methods—in order to allay subjects' fears, the research team informed subjects of the research goals in ways that may have biased their answers. The size of the interview sample may have compensated for bias, but ethnography also depends on the research team's ability to sense if informants lie or prevaricate. We omitted such interviews from our analysis. Once PLUMs were installed the time and date of installation was logged in a field notebook. At the final study site, on the day the PLUM was removed, interviewees were questioned about their toilet use habits of the day before. It was only after extensive fieldwork that we felt confident that (a) we could install PLUMs even if we asked about individual usage and (b) that asking would not bias PLUM data beyond expected reactivity.

The research team lived in the GPs while the research was conducted. This facilitated unstructured participant observation events in the form of multiple, informal visits to households to observe household sanitation practices and to triangulate interviews and PLUM data. We also assembled participant households' photographic data sets of toilet type, cabin construction, PLUM installation, and path to toilet from house. Fieldnotes on unstructured participant observation and interview transcripts were coded by recurring themes and analyzed for significant patterns. Household socioeconomic data were entered into a spreadsheet. The photographic record was organized by household and referred back to during the iterative analytical process described in the discussion section. Key informant interviews were used to create a history of sanitation interventions for each study site. After the first round of PLUM data analysis, the research team returned to the field during September 2013 for results' dissemination with stakeholders. We now turn to results and a discussion of findings from each method and as part of an iterative process.

## 13.6 Results

### 13.6.1 Qualitative Results

In brief, successful sanitation depended on three factors: political will, political ecology, and proximate social pressure. Each forms one leg of the "toilet tripod," united by political economy—the 'seat' of the toilet tripod. Political will encompassed long-term, multi-scalar government and NGO efforts to facilitate toilet building and usage. Political ecology included the complex human-environment relationships that changed over time to support toilet adoption. Proximate social pressure comprised the informal encounters that influenced neighbors and family members to build and use toilets. All four study sites had different economies, types of government intervention, NGO involvement, and environmental resources. Nevertheless, the framework of the toilet tripod comprehended the success of sanitation in each location. Below we address specific behavior, values and patterns that emerged through combining ethnography and sensor monitoring (O'Reilly and Louis 2014).

### 13.6.2   Quantitative Results

Of the 291 household data sets, a total of 258 households' data were included in the analysis. These households had PLUM readings for at least 6 days. 33 households were excluded for having less than 6 days of data, usually due to PLUM failure, and occasionally because households covered or removed PLUMs. A specialized R code for this study parsed interpreted sensor data for each household deployment across the four sites. For each sensor, outliers were removed based on 1.5 times the interquartile range for that data set, a standard outlier removal approach (Weinberg and Abramowitz 2002). For per person usage calculations, the algorithm relied on recorded household toilet user data. Children too young to use a toilet were not counted, as their feces were not generally disposed of in IHLs (O'Reilly and Louis 2014).

The data sets at each site were not normally distributed, likely due to clustered low-end recorded behavior. Therefore, groups were compared using the Wilcox ranked sum test that is less sensitive to non-normal data than the t-test. The Wilcox ranked sum difference may be interpreted as a comparable mean difference value as often presented in a t-test. Figure 13.1 and Table 13.2 show the mean per capita usage events at each of the four sites.

According to Clasen et al. (2012), a 3 min separation between usage events was arbitrarily chosen for the algorithm. We repeated this 3 min separation between usage events. If separate usage events occurred within less than 3 min of each other, the algorithm would analyze them as one usage event. Thus, underreporting during high traffic times may occur with the current analytical algorithm.

Across all four study sites, usage frequency per capita per day averaged 1.51, which is in keeping with norms for Western and non-Western populations (Palit et al. 2012). There was a slightly significant difference between WB1 (1.14) and WB2 (1.46), of about 0.245 uses per person per day. Between the two states, there was slight significance to WB (1.29) and HP (1.71) of about 0.34 uses per person per day. No statistically significant differences in per capita usage events by study site were recorded with the exception of the two sites within HP. The influence of the high per-capita toilet use in HP1 likely influenced both the state differences *and* the intra-HP differences.

## 13.7   Discussion

In this section, we discuss the insights on mean per capita usage, toilet type, and time of day of usage gained by using combined quantitative and qualitative methods.

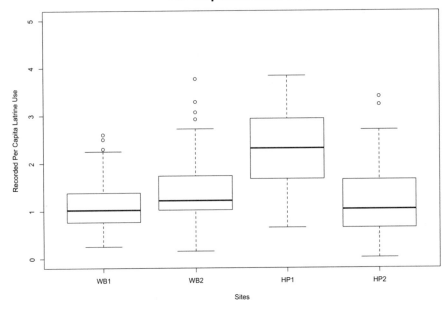

**Fig. 13.1** Per capita latrine use per day by GP

**Table 13.2** Mean per capita per day latrine use

| GP | Recorded per capita use | Wilcox ranked sum difference |
|---|---|---|
| West Bengal | 1.29 | |
| WB1 | 1.14 | 0.25 |
| WB2 | 1.46 | |
| Himachal Pradesh | 1.71 | |
| HP1 | 2.27 | 1.13 |
| HP2 | 1.18 | |
| Overall average | 1.51 | |

## 13.7.1 Mean Per Capita Usage

Initially, the data analysis suggested that WB2 per capita toilet usage was lower than WB1, but interviews led us to expect that WB2 toilet use should have been the same or higher. In WB2 the majority of households owned toilets for more than 10 years, while in WB1 the majority owned toilets for less than 10 years (see Fig. 13.2). Length of time of sustained intervention and toilet ownership meant that WB2 informants were more likely than those in WB1 to speak in terms of having a 'toilet habit.' We recalculated PLUM installations using fractions of days (as recorded in fieldnotes) to get a more accurate per capita reading than the initial calculation that

**Fig. 13.2** Histogram of toilet ownership in WB

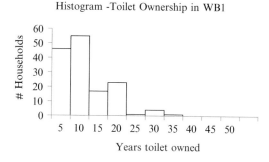

Histogram -Toilet Ownership in WB1

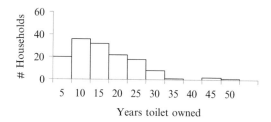

Histogram -Toilet Ownership in WB2

used whole numbers for days reported. With this adjustment, WB2 (1.46) per capita use was higher than WB1 (1.14)—a slight significant difference. Ethnography alerted us to subtleties in reported toilet usage within NGP villages, and the discrepancy between partial days and full days of installation for PLUM analysis.

The differences in mean per capita toilet usage between WB and HP were expected. In WB1 and WB2, toilets were only used for defecation and bathing after defecation. This was due to the ritual impurity of the toilet cabin, we were told, necessitating bathing and changing one's clothing after defecating inside the cabin. Urination took place outside in the family compound or nearby jungle. Family compounds nearly always had a pond, so most members bathed in the pond. For modesty's sake, some women would wash in the cabin itself. As this woman explained her reason for needing a taller, brick and mortar toilet cabin, "*My daughter cannot stand in the cabin and change her clothes now. People passing by will watch. Is this not a problem? She has to come with wet clothes inside the house.*" Previous research has noted the ways in which beliefs about impurity/disgust around feces in the South Asian context (Srinivas 2002). Our ethnography brings to light a geographically-specific, toilet-using behavior related to ritual impurity beliefs.

Using PLUM data to calculate 'total time in toilet,' HP recorded about 32 % more movement in a toilet on average than WB. This was consistent with our ethnographic research indicating that HP households use their toilet/bathrooms for other hygiene activities besides defecation. HP respondents did not report that toilet cabins were ritually impure. Instead, IHLs in both HP study sites were often built to take advantage of the single tap in family compounds, serving several purposes: toilet; bathroom; water filling station; and laundry. These larger rooms with easy access to water meant there was more traffic in and out of them, especially by women, for whom gender norms required them to do these tasks.

The differences in mean per capita usage between HP1 and HP2 were also expected. In HP1, 65 % of PLUM-accepting households had toilet/bathroom combinations. In HP2, only 23 % had toilet/bathroom combinations. When comparing usage events between toilet and toilet/bathrooms across all sites there was a significant difference (p value .00003) indicating that toilet type is important data when using PLUM technology. The difference in per capita toilet use based on toilet type indicated 0.6 fewer uses if the toilet type was 'toilet only'—validating our observations that participants spent less time in these toilet types.

We asked household members in HP1 (our last study site) on the day we removed their PLUM to recall the number of times they defecated the previous day. There was a significant difference between the sensor recorded use average of 2.27 uses per person per day, and the reported use of 1.38 for a Wilcox ranked sum mean difference of 0.85 uses. One sensor monitoring weakness is that it does not detect if the IHL is being used for the deposition of human feces. Ethnography supplied an explanation for the difference: HP1 had more toilet/bathrooms and women reported accessing stored water in the toilet/bathroom space multiple times daily. The photographic record verified that the PLUMs were installed close to toilets, but they were likely capturing non-usage events as well as usage events.

### 13.7.2 Toilet Type

We disaggregated PLUM data based on toilet quality in WB: (1) cement pan in cement slab; or (2) porcelain pan in cement slab using the photographic data set and interview data to determine whether lower cost toilets were used less than higher cost ones. Differences in toilet quality showed no significant difference in per capita usage in WB, where most low cost toilets were located across the four study areas. This result agreed with WB interviews; householders reported that low cost toilets were acceptable and in use. Using Barnard et al.'s (2013) criteria for 'functional latrine' (i.e., walls over 1.5 m; door; unbroken, unblocked pan; and functioning connection to pit (if any)), in WB, latrines were functional, even if those latrines had only plastic sheeting for walls and a door, no roof, and a cement pan. If feces could be flushed, these low cost latrines were used; this was verified by PLUM data. This key finding indicates that basic, low cost models that function are acceptable in communities where toilet use is the social norm.

In West Bengal, a GP had to achieve 90 % toilet coverage to win an NGP award. At the time that the NGP toilet drive started in the two study areas, a majority of the households could not afford to build toilets on their own. Availability of low cost cement slabs (250 INR, approximately US$5), free or subsidized pit digging, and walls of plastic sheeting supported widespread, rapid building. In WB2, 50–55 % of the households were still using cement pans. In WB1, 40–45 % had cement pans or largely subsidized porcelain pans.

There was a clear trajectory of toilet habituation in the region as one elderly man in WB2 explained, "*Earlier people used to go for open defecation OD, then khata paikhana* (pit latrine, wooden slab) *was built, then plate* (pour flush to pit latrine, cement pan) *came into existence. Now as people are making money, they are building sanitary paikhana* (pour flush to pit latrine, porcelain pan)" As his brief history relates, a significant factor in getting people to stop defecating in the open was enabling them to build pour flush latrines, even those considered temporary, as cement pan latrines were. 'Plate' latrines were a great improvement over pit latrines with wooden slabs or having to practice open defecation. Low cost latrines were less than ideal because they needed periodic reconstruction of toilet cabins, high water tables meant shallow pits (usually 3–4 rings deep) needed to be re-dug, composted, or emptied, but they did not stink, as drop pit toilets did (see also (Barnard et al. 2013; Kvarnstroem et al. 2011)). Families in WB that could afford better toilets built with porcelain pans and brick walls built them, but for those who could not, '*plate*' latrines were acceptable and were still in use decades after being built.

Pit latrines in HP were larger and had the advantage of well-draining soils and a low water table; few families had ever emptied their pits. Most latrines had porcelain pans with a cement slabs, and many families spent disposable income on tanks with piped water supply, decorative tiles, and occasionally, toilet seats.

## 13.7.3  Peak Usage Times and Occupation

PLUM data verified our ethnographic finding that most household members primarily defecated in the morning (Fig. 13.3). Data also showed a smaller but distinct peak in the evening hours. Sensors do not detect who is using the unit, a problem for per capita usage figures if household numbers fluctuate daily, but the reason households consented to installation. Using ethnography to establish family members' out-of-house routines can narrow the range of individual users throughout the day. For example, men in WB who worked as cycle-cab drivers left their houses early in the mornings and reported defecating elsewhere. Eliminating members of certain occupations as toilet users during peak hours could give more accurate mean per capita usage figures. Information on peak usage times can also assist with: knowing when to station structured observation in future studies verifying toilet usage (e.g., HP peak times were later in the morning than WB peak times (Clasen et al. 2012); capturing off-peak, high usage times (e.g., incidences of diarrhea); and informing shared toilet policy by providing information on peak time, mean per capita per hour figures (i.e., 'turnover rates').

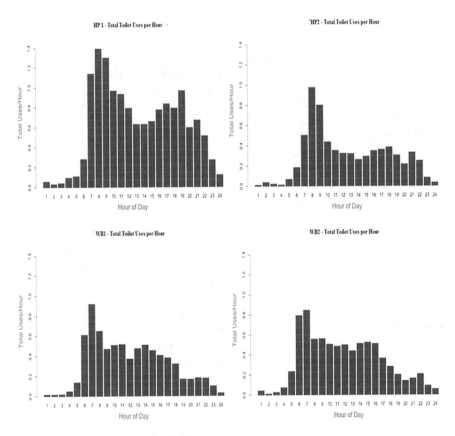

**Fig. 13.3** Time of day usage for all GPs

## 13.8 Conclusions

A failure to understand sanitation behavior can result in policies that do not meet the needs of target populations. Given high rates of open defecation in India and recently revitalized efforts to end the practice, more research is needed that measures toilet usage and explains the reasons for use and non-use. We purposefully selected unique cases to study successful sanitation uptake, intending our findings to provide new insights, guide further research, and inform interventions. We used ethnography to 'get at' the everyday lived context of study populations' toilet practices by asking people about their values, meanings, and routines. PLUMs counted 'practices,' validated interviewees' reporting, and highlighted the significance of specific behaviors.

Our mixed method approach facilitated the general findings that political will, political ecology, and social pressure supported the building and sustained usage of toilets in the study sites. Specifically, subsidies were necessary for poor households in WB to build, but these subsidized, low cost toilets were still in use decades after they were built. Contrary to findings that Indians believe latrines are expensive (Coffey et al. 2014), or that pit latrines are not sustainable (Kvarnstroem et al. 2011), low cost, improved sanitation was used sustainably. We attribute their sustainability to local governments and NGOs in WB that invested in educating families how to manage pit latrines after they filled. As Barnard et al. (2013) also found, length of time of ownership mattered for toilet use; users spoke of developing a 'toilet habit' that both supported, and was supported by, social norms in the study areas.

PLUM analysis brought to light our finding that in rural WB toilets were used only for defecation. Due to our immersion in WB, using toilets only for defecating became normalized. In seeking to explain the differences in mean per capita usage based on PLUM results, we re-discovered WB beliefs of pollution that limited toilet use to defecation. Without the ethnography we could not have explained the PLUM results for WB; without the PLUMs, defecation-only toilet use would have been overlooked. An understanding that a toilet cabin is a polluting space presents new challenges for solving problems such as the disposal of child feces (Jenkins et al. 2014) or needed privacy for urination. Currently, PLUMs detect motion in and out of the toilet cabin without information on what occurred inside. Rural WB also presents itself as a place where the PLUM algorithm for 'usage events' might be further refined to assess 'defecation events' since toilets are used only for defecation. Other instruments including audio signal analysis or pressure pads placed near the toilet could also be field tested in WB as further improvement to PLUMs.

As in other studies, we found that not all family members regularly used toilets (Coffey et al. 2014; Jenkins et al. 2014) but interview data can enable refinement of PLUM data analysis by collecting information on the age and occupation of non-users. This serves the purpose of refining mean per capita usage, and thereby letting us know if the toilet is being used, by how many, and at what time. Standard large-scale survey methods could provide some of the same data (Barnard et al. 2013; Jenkins et al. 2014) and be verified by sensor monitoring, but without knowledge of norms and meanings, solutions to problems of non-usage due to occupation and age remain out of reach.

Ethnography relies on trust between the research team and the study community, not just individual interviewees. In small villages in WB and HP occupied by extended families, a misstep could have ended our research at those sites. The question of trust when using combined methodology raises the question as to whether people would be willing to install if they did not live in NGP villages? As stated above, we learned early on that PLUM installations were possible when households were informed that we chose their GP because it was an NGP village—because we knew their toilets were in use. Given the difficulty of installation in places of successful sanitation, installation in locations where populations were informed that they should use toilets but did not, would likely have low PLUM acceptance and could undermine the trust necessary for a rich ethnography.

Ethnography is seldom undertaken as it requires extended field periods and linguistic and cultural fluency, but its strengths lie in discovering new practices, and the surprising, subtle motivations for behaviors. Such discoveries are critical in their own right, but they also can inform other assessment tools. Findings can only be scaled up with caution, because 'scaling up' requires removing norms and meanings from the geographic context where they arose—in this case, tantamount to ignoring the very multi-scalar and intersecting factors (e.g., governance, changing environmental conditions, and processes of social norm development) that produced the conditions of successful sanitation. Similarly, PLUMs are not appropriate for wide-scale measurement of toilet usage in India, given the diversity of behaviors and beliefs across small geographic areas. Nevertheless, the findings from our combined methodology indicate that ethnography and sensor monitoring are important tools in the search for methods to assess toilet usage and behavior.

# References

Barnard S, Routray P, Majorin F, Peletz R, Boisson S, Sinha A, Clasen T (2013) Impact of Indian total sanitation campaign on latrine coverage and use: a cross-sectional study in Orissa three years following programme implementation. PLoS One 8(8):e71438. doi:10.1371/journal. pone.0071438

Clasen T, Fabini D, Boisson S, Taneja J, Song J, Aichinger E, Bui A, Dadashi S, Schmidt WP, Burt Z, Nelson KL (2012) Making sanitation count: developing and testing a device for assessing latrine use in low-income settings. Environ Sci Technol 46(6):3295–3303. doi:10.1021/es2036702

Clasen T, Boisson S, Routray P, Torondel B, Bell M, Cumming O, Ensink J, Freeman M, Jenkins M, Odagiri M, Ray S, Sinha A, Suar M, Schmidt WP (2014) Effectiveness of a rural sanitation programme on diarrhoea, soil-transmitted helminth infection, and child malnutrition in Odisha, India: a cluster-randomised trial. Lancet Glob Health 2(11):e645–e653. doi:10.1016/S2214-109X(14)70307-9

Coffey D, Grupta A, Hathi P, Nidhi K, Spears D, Sriyastav N, Vyas S (2014) Revealed preference for open defecation: evidence from a new survey in rural North India. Econ Pol Week 49(38):43–55

Cousens S, Kanki B, Toure S, Diallo I, Curtis V (1996) Reactivity and repeatability of hygiene behaviour: structured observations from Burkina Faso. Soc Sci Med 43(9):1299–1308

Delea MG, Nagel C, Thomas E, Halder AK, Amin N, Shoab AKM, White Z, Freeman MC, Unicomb L, Clasen TF. Assessing latrine utilisation: A comparative analysis of respondent-reported and instrument-recorded utilization in rural Bangladesh. 2015 (unpublished)

Drangert JO, Nawab B (2011) A cultural-spatial analysis of excreting, recirculation of human excreta and health–the case of North West Frontier Province, Pakistan. Health Place 17(1):57–66. doi:10.1016/j.healthplace.2010.08.012

Jenkins M (2004) Who buys latrines, where and why? Water and sanitation program, field note, September 2004. The World Bank, Washington, DC

Jenkins MW, Curtis V (2005) Achieving the 'good life': why some people want latrines in rural Benin. Soc Sci Med 61(11):2446–2459. doi:10.1016/j.socscimed.2005.04.036

Jenkins MW, Freeman MC, Routray P (2014) Measuring the safety of excreta disposal behavior in India with the new Safe San Index: reliability, validity and utility. Int J Environ Res Public Health 11(8):8319–8346. doi:10.3390/ijerph110808319

Kvarnstroem E, McConville J, Bracken P, Johansson M, Fogde M (2011) The sanitation ladder – a need for a revamp? J Water Sanit Hyg Dev 1(1):3–12

Manun'Ebo M, Cousens S, Haggerty P, Kalengaie M, Ashworth A, Kirkwood B (1997) Measuring hygiene practices: a comparison of questionnaires with direct observations in rural Zaire. Trop Med Int Health 2(11):1015–1021

McFarlane C (2008) Sanitation in Mumbai's informal settlements: state, 'slum', and infrastructure. Environ Plan A 40(1):88–107

Montgomery MA, Desai MM, Elimelech M (2010) Assessment of latrine use and quality and association with risk of trachoma in rural Tanzania. Trans R Soc Trop Med Hyg 104(4):283–289. doi:10.1016/j.trstmh.2009.10.009

O'Reilly K (2010) Combining sanitation and women's participation in water supply: an example from Rajasthan. Dev Pract 20(1):45–56

O'Reilly K, Louis E (2014) The Toilet Tripod: understanding successful toilet adoption in rural India. Health Place 29:43–51

Olsen A, Samuelsen H, Onyango-Ouma W (2001) A study of risk factors for intestinal helminth infections using epidemiological and anthropological approaches. J Biosoc Sci 33(4):569–584

Palit S, Lunniss PJ, Scott SM (2012) The physiology of human defecation. Dig Dis Sci 57(6):1445–1464. doi:10.1007/s10620-012-2071-1

Ram PK, Halder AK, Granger SP, Jones T, Hall P, Hitchcock D, Wright R, Nygren B, Islam MS, Molyneaux JW, Luby SP (2010) Is structured observation a valid technique to measure hand-washing behavior? Use of acceleration sensors embedded in soap to assess reactivity to structured observation. Am J Trop Med Hyg 83(5):1070–1076. doi:10.4269/ajtmh.2010.09-0763

Rheinlander T, Samuelsen H, Dalsgaard A, Konradsen F (2010) Hygiene and sanitation among ethnic minorities in Northern Vietnam: does government promotion match community priorities? Soc Sci Med 71(5):994–1001. doi:10.1016/j.socscimed.2010.06.014

Rodgers AF, Ajono LA, Gyapong JO, Hagan M, Emerson PM (2007) Characteristics of latrine promotion participants and non-participants; inspection of latrines; and perceptions of household latrines in Northern Ghana. Trop Med Int Health 12(6):772–782. doi:10.1111/j.1365-3156.2007.01848.x

Small ML (2009) 'How many cases do I need?' On science and the logic of case selection in field-based research. Ethonography 10(1):5–37

Srinivas T (2002) Flush with success: bathing, defecation, worship, and social change in South India. Space Cult 5(4):368–386

Thomas E, Zumr Z, Graf J, Wick C, McCellan J, Imam Z, Barstow C, Spiller K, Fleming M (2013) Remotely accessible instrumented monitoring of global development programs: technology development and validation. Sustainability 5(8):3288–3301

Thomson P, Hope J, Foster T (2012) GSM-enabled remote monitoring of rural handpumps: a proof-of-concept study. J Hydroinf 14(4):29–39

Weinberg M, Abramowitz SK (2002) Data analysis for the behavioral sciences using SPSS. Cambridge University Press, Cambridge/New York

# Chapter 14
# Sustainable Sanitation Provision in Urban Slums – The Sanergy Case Study

**David Auerbach**

**Abstract** Sanergy, a Nairobi-based social enterprise, tackles the sanitation crisis in urban informal settlements in Africa. We take an innovative, systems-based approach that addresses the entire sanitation value chain. We build high-quality, low-cost sanitation units, known as Fresh Life Toilets, which we franchise to community members, who run them as businesses. We collect the waste on a regular basis, removing it from the community. We then convert the waste into valuable by-products, including organic fertilizer and insect-based animal feed, which we sell to regional farmers. Through this model, we are making it profitable – and thus sustainable – to provide hygienic sanitation in urban slums.

**Keywords** Sanergy • Sanitation service • Value chain

## 14.1 Introduction

More than four billion people in the developing world lack access to total hygienic sanitation (Baum et al. 2013). The consequences are staggering: 760,000 children under the age of five die each year from diarrheal diseases, 90 % of which are due to poor sanitation (WHO 2013). Globally, $260 billion is lost in diminished productivity and healthcare costs. On average, inadequate sanitation costs countries 1.5 % of their GDP. In Kenya, where my company Sanergy has worked since 2011, the costs of poor sanitation are $324 million per year, with $51 million going toward medical and health care costs (World Bank 2013).

While improving sanitation isn't as attractive a cause as other development problems – clean water, education, and income generation, for example – sanitation

D. Auerbach (✉)
Sanergy Limited, Theta Lane, Kilimani, Nairobi, Kenya
e-mail: david@saner.gy

© Springer International Publishing Switzerland 2016
E.A. Thomas (ed.), *Broken Pumps and Promises*,
DOI 10.1007/978-3-319-28643-3_14

211

is a root cause of many development issues. Clean water means that the water is free from contamination, mostly due to fecal matter. Properly removing and treating waste is key to ensuring water is and remains clean and potable. Children often miss school, falling behind in their studies, because of sanitation-related illnesses, and adolescent girls frequently drop out of school or fall drastically behind when they start their periods because they prefer to stay home when menstruating instead of dealing with the indignity of using inadequate school facilities. And sanitation-related illnesses often impact adults' ability to hold jobs or otherwise earn a living to support themselves and their families. Because of these ramifications and more, every $1 invested in sanitation sees a fivefold return (World Bank 2013).

Millennium Development Goal 7(c), which aimed to halve the number of people without sustainable access to sanitation by the end of 2015, was not met, representing one of the greatest failures of the MDGs. The Sustainable Development Goals, which the United Nations approved in September 2015, build on the MDGs to commit member countries to achieve further progress by 2030. The SDGs have revised the MDG commitment to "achieve access to adequate and equitable sanitation and hygiene for all and end open defecation" by 2030 (UN 2015). In order to achieve this, there is still much work to do.

A lack of basic infrastructure makes the sanitation crisis particularly acute in urban slums, where populations will double to two billion in the next 15 years (UN-Habitat 2003). In Kenya, eight million slum residents still have to resort to unhygienic and undignified sanitation solutions, such as "flying toilets" (defecating into plastic bags that are then tossed onto the streets) and pay-per-use pit latrines that release untreated human waste into the environment (O'Keefe et al. 2015).

Only 10–15 % of Nairobi's slums are sewered, and sewer pipes are often broken or clogged. In total, four million tons of human waste from Kenya's slums are dumped untreated into waterways each year – polluting the environment, spreading disease, and harming community health. At current rates, reaching complete sanitation coverage in Kenya will take 150 years. The loss of productivity due to sanitation-related illness costs Kenya's GDP a million dollars a day (O'Keefe et al. 2015; UN-Habitat 2006).

In the case of Sanergy, solving the sanitation crisis requires more than just building toilets. Sanergy's innovation is to take a systems-based approach that engages the community at every step and, in doing so, guarantees that residents of slums gain access to the hygienic sanitation services they both need and want.

First, we build high-quality, low-cost "Fresh Life Toilets." They are designed with qualities users desire: they are easy to keep clean and maintain; their small footprint (1 m by 1.5 m) allows them to be installed close to homes; and they include hand-washing stations to promote good hygiene practices. Underneath the toilet, easy-to-remove cartridges capture the waste, ensuring it does not pollute the soil and waterways.

Then, we franchise Fresh Life Toilets (FLTs) to local residents in Nairobi's informal settlements through three models: commercial, residential, and in community institutions, such as schools. The owners – Fresh Life Operators (FLOs) – invest to become franchise partners, putting skin in the game and creating accountability for both the operator and Sanergy.

We provide FLOs with access to interest-free financing, help in securing land access, business training, aspirational marketing, ongoing operational support, and guaranteed waste collection service. The operator commits to cleaning the toilets, keeping them consistently open, and generating demand using his or her local credibility and influence. Through frequent field visits and spot inspections, we ensure that FLTs across the network are maintained to the same standards of cleanliness and hygiene. In this way, community members contribute to the health of their neighbors – a responsibility they take very seriously. At quarterly forums, FLOs discuss successes and obstacles in providing their communities with hygienic sanitation. FLOs learn from one another, sharing best practices and advice for improved service delivery.

The Sanergy waste collection team then collects the waste from each toilet on a regular basis, replacing the full collection cartridges with clean, empty ones. Once the waste has been removed from the community, we convert the waste into a variety of saleable end products, including organic fertilizer, called Evergrow, and insect-based animal feed. These end products are then sold to Kenyan farmers, who see a 30 % increase in their crop yields and restored soil health when they use Evergrow.

## 14.1.1 Making Sanitation Provision Profitable

In order for Sanergy to successfully provide hygienic sanitation at scale, it is imperative that the community be invested in making our model work. In order to incentivize community buy-in, we ensure that community residents recognize the benefits of their participation – be it financial, health, or environmental.

Unemployment in the informal settlements in which we work is about 40 % – and even higher among youth populations (UN-Habitat 2012). This causes huge social problems, including crime and violence from unemployed, unskilled youth. We make an effort to fill as many jobs as possible from the communities we serve, and almost half our staff members come from Nairobi's informal settlements. We help with career and professional development, so that our employees have the skills and training they need to succeed at Sanergy and beyond.

## 14.1.2 Incentivizing Fresh Life Operators

In order to help ensure community buy-in for its model, Sanergy distributes the majority of FLTs through a franchise model, in which local community members run and maintain the toilets, charging customers a nominal fee per use. The value proposition for potential FLOs is twofold: earn a steady income and improve the health of community residents.

Since 2012, we have had a partnership with Kiva, an online micro-lending platform, to help potential FLOs gain access to interest-free loans with which to purchase an FLT, which costs around 500 USD. Once approved for a loan, FLOs are able to choose between either a 12-month or a 24-month loan. After paying an initial down payment of about 20 %, they use revenue generated from running the toilet to pay down the balance of the loan. Our credit team services the loans, ensuring timely payments and low default rates.

The costs of running an FLT are fairly low; operators are responsible for buying toilet paper and sawdust and ensuring the handwashing station has water and soap for all customers. With an average of 75 users per day, an FLO can earn 100,000 Kenyan shillings (about 1,000 USD) per toilet per year – a solid income for residents of Nairobi's informal settlements. Most FLOs run at least two toilets, which increases their income even more.

Many FLOs also run other businesses. Hannah Muthoni, for example, has two FLTs and two showers next to a small shop, where she offers a variety of goods for her neighbors. The income she earned from her first FLT helped finance this expansion, which means she no longer has to travel a long distance to the local market to sell her goods, and she can now take care of her grandchildren while her daughters – who have more education and therefore have higher earning potential than Hannah does – are at work.

### 14.1.3  Incentivizing Sanergy

To make sanitation provision profitable, and thus sustainable, Sanergy has three main revenue streams. The first is selling toilets we manufacture with local materials and labor. Our customers include corporates, NGOs, and government entities looking for a hygienic sanitation solution. Our unique urine-diverting dry toilet is a waterless hygiene solution adaptable to many locations and circumstances and does not require investment in additional infrastructure, making it an appealing solution for customers, especially in areas where there is no sewerage coverage.

We have also built a robust and efficient waste-collection network, which currently removes 9–10 tons of waste per day and works in complement with our infrastructure distribution network. In addition to the cost of the toilet, FLOs pay an annual renewal fee of about 90 USD for our waste collection services. This renewal fee is less expensive and more convenient than hiring a vacuum truck or other exhaustion service, as pit latrine owners have to do.

At a centralized facility, Sanergy converts the waste into end products for which there is high demand in the region. Through a co-composting process, Sanergy's processing team converts most of the waste into Evergrow, a nutrient-rich, pathogen-free organic fertilizer.

The Kenyan Ministry of Agriculture has identified soil degradation – due to a lack of crop rotation and the use of harsh chemical fertilizers – as the number one threat to food security in East Africa. To restore soil health, they have recommended Kenyan farmers use organic fertilizer on their crops. This recommendation is difficult for farmers to follow: there is currently little domestically produced organic fertilizer, and imported fertilizers are often prohibitively expensive. Sanergy's leadership saw a market opportunity for valuable end products, specifically Evergrow.

Using waste as a fertilizer is a common practice throughout the developed world – including the United Kingdom and the United States. In fact, about 60 % of all treated sewage sludge in the U.S. is applied to fields, where the nitrogen and phosphorous in the sludge helps crops grow (George 2008). Evergrow has been shown to restore soil health and increase crop yields by 30 %.

In addition, Sanergy has been trialing the development of insect-based animal feed derived from Black Soldier Flies (BSF), the larvae of which feed on organic waste. Once the larvae stop feeding, they are boiled and dried, resulting in a high-protein animal feed, suitable for a variety of livestock. The East African animal feed market is growing steadily at about 7 % per year, and livestock farmers are dissatisfied with the currently available options, both because of quality and inconsistent supply. Sanergy's trials have gone well thus far, and the BSF operations are expanding rapidly.

In partnership with a variety of organizations, including the Bill & Melinda Gates Foundation and Reinvent The Toilet, Sanergy is also trialing several other end products, including biogas, liquid fertilizer made through urine valorization, and biochar to be used as a soil amendment. This diversification of end products enables Sanergy to cater to a wide customer base and address the wide array of needs East African farmers have. Sanergy's R&D is primarily funded through grant capital, allowing for experimentation to ensure we can develop end products that efficiently convert waste into something of value for our customers.

The production of Evergrow fertilizer and other end products relies on the waste collected from Fresh Life Toilets each day. The more sanitation services we offer, the more end products we can make. The more end products we sell, the easier and more widespread it becomes to provide sanitation affordably. This is how, in working to address the sanitation challenge, we are also tackling East Africa's agricultural productivity crisis.

As a young company, Sanergy is not yet profitable; however, we are confident we have developed an economically viable model that will allow the sustainable provision of hygienic sanitation in urban informal settlements. We are working to scale the model to reach profitability, in addition to achieving maximum impact.

## 14.1.4  Next Steps

The results of the Sanergy model so far are promising. In just 4 years, we have launched 781 Fresh Life Toilets in Nairobi's slums, run by 387 operators. The network of Fresh Life Toilets is used over 33,000 times per day. Sanergy ensures the removal of 60 tons of waste from the toilets per week, and the waste is converted into end products that help Kenyan farmers increase their crop yields and keep their animals well-fed.

Looking forward, we are committed to achieving 100 % coverage in the areas we serve. We have already learned many lessons from our recent expansion into the slum of Mathare, especially about the parts of our model that need to adapt to hyperlocal contexts. The Sanergy model relies on community buy-in, and residents need to understand why this is the best way to ensure the health and prosperity of their families and friends. We work closely with residents to tailor our offerings to the needs and desires of our customers, so that we can be confident that people are willing to pay for Sanergy's services.

Commercial operators running two FLTs generate about 2,000 USD per year in profit from charging a minimal usage fee to customers. Schools have seen significant increases in attendance and enrolment after installing FLTs, and in residential compounds, plot owners have seen occupancy go up by 60 %, and more timely rent payments. The message is clear: if we can provide the services they demand, residents of slums will invest in hygienic sanitation.

## References

Baum R, Luh J, Bartram J (2013) Sanitation: a global estimate of sewerage connections without treatment and the resulting impact on MDG progress. Environmental Science and Technology 47(4):1994–2000

Delea MG, Nagel C, Thomas E, Halder AK, Amin N, Shoab AKM, White Z, Freeman MC, Unicomb L, Clasen TF (2015) Assessing latrine utilisation: A comparative analysis of respondent-reported and instrument-recorded utilization in rural Bangladesh (unpublished)

George R (2008) Too good to waste? The Guardian. http://www.theguardian.com/environment/2008/aug/29/waste.recycling. Accessed 28 Aug 2008

O'Keefe M, Lüthi C, Tumwebaze IK, Tobias R (2015) Opportunities and limits to market-driven sanitation services: evidence from urban informal settlements in East Africa. Environ Urban 27(2):421–422–440

UN (2015) Goal 6: ensure access to water and sanitation for all United Nations. New York. http://www.un.org/sustainabledevelopment/water-and-sanitation/. Accessed 15 October 2015.

UN-Habitat (2003) The challenge of slums – global report on human settlements. United Nations, Nairobi

UN-Habitat State of the Urban Youth 2012/2013 (2012) In Nairobi

UN-Habitat State of the World's Cities 2006/2007 (2006) In Nairobi

WHO (2013) Fact Sheet Number 330: Diarrhoeal disease. http://www.who.int/mediacentre/factsheets/fs330/en/. Accessed 14 October 2015

World Bank (2013) What's a toilet worth? World Bank, Washinton, DC

# Chapter 15
# Pay for Performance Energy Access Markets

**Dexter Gauntlett, Michael Ronan Nique, and Helene Smertnik**

**Abstract** Few examples of pay-for-performance service delivery models have been as successful, particularly in the development context, as the "Energy Access" market. Originally the domain of non-governmental organizations and government rural electrification programs in developing countries, the energy access sector has transitioned to a vibrant marketplace with potential for continued scaling. The market has developed to the point, enabled by rapid reductions in solar PV and lighting prices, expanded telecom coverage, innovation in machine-to-machine technology development, and "last-mile" business models, that traditional aid now is seen as a threat to further market growth. This article examines the trends that enabled the rapid deployment of off-grid solar markets based on pay-as-you-go electricity service that is yielding significant health and economic benefits for those living in many of the most remote communities in developing countries.

**Keywords** Solar lighting • Solar lanterns • PAYG • Energy access

## 15.1 Introduction

Up to 1.4 billion people worldwide, including nearly 600 million in Sub-Saharan Africa and 800 million in Asia, are without access to electricity (IEA 2010b). These populations previously had no choice but to pay high prices for low-quality and

---

Sections 15.1, 15.2, 15.3, 15.4, 15.4.1, 15.4.2, 15.4.2.1, 15.4.2.2, 15.4.2.3, 15.4.2.4 and 15.4.3 were written by Dexter Gauntlet, unless otherwise attributed, and are based on his research that first appeared in the Navigant Research report entitled *Solar PV Consumer Products*, dated May 2014, and used by permission of Navigant Consulting, Inc. Remaining sections adapted with kind permission from Springer Science + Business Media: *Decentralized Solutions for Developing Economies, The Synergies Between Mobile Phone Access and Off Grid Energy Solutions.* 2015.

D. Gauntlett
Navigant Research, 1320 Pearl Street, Suite 300, Boulder, CO 80302, USA

M.R. Nique (✉) • H. Smertnik
GSM Association, GSMA, 2nd Floor The Walbrook Building,
25 Walbrook, London EC4N8AF, UK
e-mail: mnique@gsma.com

© Springer International Publishing Switzerland 2016                                217
E.A. Thomas (ed.), *Broken Pumps and Promises*,
DOI 10.1007/978-3-319-28643-3_15

polluting fuel-based lighting such as kerosene lamps. Kerosene costs as much as 50 % more in remote areas compared to urban areas due to transport challenges, further contributing to the cycle of poverty. In addition to providing inadequate illumination, kerosene lamps pose significant health risks. New advancements in lighting technology have enabled the development of pico solar systems, which are compact, clean, and affordable off-grid lighting and energy products. Many of these products use solar charging (<10 W) and light-emitting diode (LED) lighting technology. As with most renewable energy technologies, solar lighting is typically more affordable compared to conventional lighting primarily from kerosene, but upfront capital costs (even if only $10) can be a challenge to last-mile customers.

The market for off-grid solar lighting has reached a point where there is considerable opportunity around the world and multiple entry points for manufacturers, distributors, service providers, and others throughout the value chain. Platform companies have been established, investment is flowing, and companies are expanding, so opportunity is abounding. However, so is risk and uncertainty as the industry turns from aspirational to big business.

While technology cost reductions and strong financial support from aid organizations initially combined to jumpstart the pico solar market – solar microcredit schemes and innovative pay-as-you-go (PAYG) machine-to-machine (M2M) integrated solar technologies in base of the pyramid (BOP) markets have enabled the vibrant market that exists today. Together, these trends are reducing barriers to ownership of solar lighting for rural customers, improving health, and enabling a pathway out of poverty.

## 15.2 Technology Definition

While the off-grid solar lighting industry today lacks an official common set of definitions or terminology for solar lighting products, the following Lighting Asia (World Bank 2012) definitions are generally accepted for BOP markets:

- **Flashlights/torches**: Portable handheld devices offering directional lighting at low lumen output. Today's solar torches typically feature integrated solar panels.
- **Task lamps/work lights**: Portable or stationary handheld devices, including solar desk lamps, in a range of panel sizes and light output levels utilized for specific tasks (e.g., reading, weaving, etc.).
- **Ambient lamps/lanterns**: Portable or stationary devices that resemble the kerosene hurricane lamp form factor. These devices typically offer multidirectional light, along with a variety of sizes and functionality depending on technology (e.g., from heavy powerful compact fluorescent lamps [CFLs] to smaller LED-based systems).
- **Multifunctional devices**: Portable or stationary devices that can provide directional and multidirectional light, feature a variety of value-added features

(e.g., mobile phone charging), and can be utilized for either task-based or ambient lighting needs.

- **Micro-SHSs**: Semiportable lighting devices associated with a small portable solar panel that powers or charges one to three small lights, mobile phones, and other low-power accessories (e.g., radio, mini fan). These products are distributed across a spectrum of decreasing specialization and increasing lumen output, which correlates directly with the increasing capacity of the solar panel typically attached to them.

These products can be grouped into two key markets – pico solar and solar home systems (SHS).

- **Pico solar systems**: These are standalone lighting appliances that cost between $10 and $100 and integrate small solar modules (generally <10 W) with white LEDs and storage batteries. Products frequently also include additional capabilities such as mobile phone charging. The vast majority of pico solar systems are sold in the developing world, though several pico solar products in the United States and Europe are gaining traction in the outdoor segments (e.g., solar lanterns for camping, solar coolers, and solar-integrated backpacks for battery charging).
- **SHSs**: A typical SHS includes a solar panel (10 W–200 W), a vehicle lead-acid storage battery, DC-powered lights (including white LEDs), and often times DC-Powered appliances such as a radio and/or television. The entire system costs up to a few hundred dollars. SHSs are exclusively sold in the developing world and serve as the primary source of electricity for base of the pyramid (BOP) consumers. Non-governmental organizations (NGOs), social enterprises, and government aid agencies distribute, sell, and finance these systems as part of energy access initiatives.

Beyond solar PV consumer products are community-scale systems, often referred to as microgrids.

- **Microgrids**: Microgrids for rural electrification are based on generation (frequently solar PV) of a few kilowatts (kW) to tens of kW, often paired with battery storage, though there is significant variation in size and technologies (Table 15.1).

## 15.3   History and Key Locations for Pico Solar and SHS Markets

In the 1990s and early 2000s, government-led rural electrification programs, such as those in India, Bangladesh, Brazil, and Morocco, experimented with deploying SHSs at scale and paved the way for new solar energy service companies to enter the market today. In all cases, SHSs were identified as more cost-effective than extending the grid to the most remote communities. The primary challenge for all of

**Table 15.1** Off-grid rural electrification segments

| Metric | Pico solar systems | SHSs | Microgrids |
|---|---|---|---|
| Relative system size | <10 W | 10 W–200 W | 1 kW–20 kW |
| Typical applications | Lighting, phone charging | Lighting, phone charging, fans, radios, small appliances | All household and business electricity use, refrigeration, basic machinery, water pumping |
| PAYG enabled? | Yes | Yes | Yes |
| Indicative product distribution channels | Social enterprises, NGOs, government, retail | Social enterprises, NGOs, government | Energy and infrastructure developers |
| Typical target customer | Individuals | Individuals | Rural communities |

these programs was overcoming the high upfront costs of solar PV systems. The programs that were most successful were those that enabled customers to pay for the system over a number of years. Governments quickly learned the importance of a grassroots approach to distributing SHSs and providing access to microcredit. They eventually also recognized the need to establish strong after-sales service networks. In the 1990s and early 2000s, SHSs were the only option on the market, as the pico solar lighting industry had yet to emerge.

Successful implementation of many of these government-led programs led to follow-on programs, such as those in Bangladesh, India, and Nepal, that sought to continue to scale up their programs through traditional aid mechanisms. The humanitarian/NGO community often became distributors of SHSs that were paid for by donor money. As incomes continued to rise in these markets and the concept of microcredit was well-proven in terms of high payback rates, the market shifted to a more private-sector approach. Subsidies were and still are an important aspect of some markets, but customer willingness to pay opened the door to new types of products such as pico solar lanterns – and the removal of subsidies in new markets. New companies emerged, manufacturing low-cost lighting solutions that would be affordable to rural customers and using the same grassroots distribution approach as NGO and government implementers. At the same time, increased political pressure to reduce fuel subsidies for diesel and kerosene in particular, combined with dropping solar PV prices, more efficient lighting advances, and the advent of social enterprises, gave way to the modern pico solar lighting industry that exists today.

## 15.4 Looking Ahead

To achieve universal energy access for the 1.2 billion living in energy poverty by 2030, the International Energy Agency (IEA) estimates it will cost $33 billion a year, or a total of $700 billion between 2010 and 2030 (IEA 2010a). Data from the

IEA indicates that concentrations of unelectrified populations are (and will continue to be) in Sub-Saharan Africa, India, and in other parts of Asia-Other. Asia-Other refers to every other country in Asia except for China.

Both Sub-Saharan Africa and Asia will have more than half a billion people without electricity in 2030, with over half the unelectrified Asian population in India (293 million). Notably, the unelectrified population will continue to rise in Sub-Saharan Africa through 2030, while it will fall in other regions of the world (IEA 2010a).

As the primary source of lighting for millions of people, kerosene displacement underpins the business model for the energy access consumer market. Not only is kerosene use a major contributor to poor health outcomes, but it is commonly one of the largest sources of household expenditure.

### 15.4.1  Kerosene Displacement

According to International Finance Corp.'s (IFC's) Lighting Africa program, the African BOP population spent between $13.2 billion and $17.3 billion annually on non-renewable, fuel-based lighting (including expenditures on kerosene, batteries, and candles) in 2012. Off-grid households account for $10.5 billion to $14 billion, or about 80 % of the total lighting expenditure, with spending by under-electrified households and small and medium enterprises (SMEs) accounting for the remaining 20 %. Kerosene remains the most important lighting fuel for off-grid and under-electrified households and SMEs in Africa and accounts for approximately 55 % of total BOP lighting expenditure, according to Lighting Africa. Furthermore, rural households pay a premium for kerosene of 25–175 % more compared to urban households due to logistical difficulties (IFC 2012).

Kerosene prices are also highly volatile and correlate with oil prices. Between 2000 and 2012, kerosene prices increased 240 % in the developing world, from an average price of roughly $0.50 per liter in 2000 to about $1.20 per liter in 2012. In high-cost markets, including Burundi, Guatemala, and Panama, kerosene costs can be as high as $1.80–$2.10 per liter.

Kerosene is also highly detrimental to health and the environment. Use of kerosene subjects people to multiple pollutants, including fine particulate matter, formaldehyde, carbon monoxide, polycyclic aromatic hydrocarbons, sulfur dioxide, and nitrogen oxides. Exposure to these pollutants can result in an increased risk of respiratory and cardiovascular diseases, cancer, and mortality (Lam et al. 2012). In addition, there are other acute health impacts from having kerosene in the home, including burns, fires, and ingestion of kerosene. Despite these hazards, kerosene leads the market in many African countries that rely on alternatives such as burning wood, charcoal, and dung for cooking and lighting.

## 15.4.2 Energy Access and Rural Electrification Programs Expand Market Opportunity for PAYG

Energy access has been identified as one of the key measures to increase incomes and reduce poverty. The United Nations (UN) Sustainable Energy for All program is a multi-stakeholder partnership between governments, the private sector, and civil society. Launched by the UN secretary-general in 2011, the program has three inter-linked objectives to be achieved by 2030: ensure universal access to modern energy services; double the global rate of improvement in energy efficiency; and double the share of renewable energy in the global energy mix. The UN has declared 2014–2024 as the Decade of Sustainable Energy for All to galvanize efforts to make universal access to sustainable modern energy services a priority. It notes that 1.3 billion people are without electricity worldwide and 2.6 billion people in developing countries rely on traditional biomass for cooking and heating. The Sustainable Energy for All initiative has complemented and, in many cases, accelerated the work by national governments to electrify rural areas. It has also led to changes in the approach to rural electrification – away from grid extension and increasingly toward off-grid lighting products.

The following sections discuss other initiatives that have led to the growth in the energy access market and/or will continue to be important drivers in the future. While not all of these programs include PAYG integrated solar technologies, it is expected that these initiatives will not only expand the opportunity for PAYG products already discussed here, but also new products and services related to energy and other sectors – such as water and sanitation.

### 15.4.2.1 Lighting Global

Lighting Global, a joint World Bank and IFC program, was originally established as Lighting Africa. The latter sought to accelerate the development of markets for modern off-grid lighting products in Sub-Saharan Africa, where an estimated 10–30 % of household income is spent on hazardous and low-quality fuel-based lighting products. Lighting Africa achieved its initial goal of mobilizing public- and private-sector support in order to supply high-quality, affordable, and safe lighting to 2.5 million people by facilitating the sale of 500,000 off-grid lighting units by 2012. At the same time, the program seeks to create a sustainable commercial platform to realize the vision of providing 250 million people with modern off-grid lighting products by 2030. Promoting the use of improved, low-cost off-grid lighting technology will provide an avenue for social, health, and economic development, especially for households and small businesses, which will realize significant cost savings and increases in productivity from the transition. The success of Lighting Africa has led to the creation of Lighting Asia and Lighting Global.

### 15.4.2.2  Infrastructure Development Company Ltd.'s Bangladesh Solar Home Systems Program

Infrastructure Development Company Ltd. (IDCOL) promotes the dissemination of SHSs in the remote areas of Bangladesh with financial support from The World Bank, Global Environment Facility, KfW, GTZ, Asian Development Bank, and Islamic Development Bank. IDCOL started the program in January 2003 and its initial target was to finance 50,000 SHSs by the end of June 2008. The target was achieved in September 2005, 3 years ahead of schedule and $2.0 million below the estimated project cost. IDCOL then revised its target and decided to finance 200,000 SHSs by the end of 2009. This was also achieved in May 2009. The company's current target is to finance 6 million SHSs by the end 2016. As of November 2013, 2,677,896 SHSs had been installed through the program via 46 partner organizations that sell SHSs (IDCL 2014).

### 15.4.2.3  Energy Sector Assistance Programme

Energy Sector Assistance Programme (ESAP) is a joint program in Nepal under the Alternative Energy Promotion Centre (AEPC) supported by the governments of Nepal, Denmark, Norway, Germany, and the United Kingdom. Since 2001, the Solar Energy Component of AEPC/ESAP has supported rural electrification through the dissemination of solar PV systems. To date, AEPC/ESAP has supported the installation of more than an estimated 400,000 SHSs in 73 districts in Nepal.

### 15.4.2.4  Other National Energy Access Initiatives

Many other developing nations have rural electricity strategies, though not all are specifically promoting the use of solar PV consumer products. This is the case in Latin America, where pico solar and SHS companies have found it difficult to successfully penetrate the market despite similar percentages of unelectrified populations in some countries. Other notable national energy access initiatives include:

- **India's National Solar Mission** seeks to achieve 20 GW of solar power by 2022, in part through the installation of rooftop PV systems. It has also set the specific goal of providing 20 million solar lighting systems in place of kerosene lamps to rural communities during this time, hoping to reach an estimated 100 million people.
- The **Ghana Solar Lantern Distribution project** is a program supported by the government of Ghana to provide subsidies to support sales of 200,000 solar lanterns between 2014 and 2016. The program uses money formerly allocated for fuel subsidies.

- **Kenya's kerosene phase-out program** was announced in mid-2012 to eliminate the use of kerosene for lighting and cooking, replacing the fuel with clean energy products. Norway has pledged $44.5 million toward the initiative.

### 15.4.3 Mobile Money Enables Energy Access Markets to Thrive via PAYG Business Models

Most of the leading organizations in the off-grid solar lighting sector position themselves as social enterprises. These for-profit or not-for-profit private entities have a social mission and use private-sector methods to generate financial returns that enable them to scale more effectively. While the solar lighting industry has started to transition to a private-sector model, the role of grants and subsidies still plays a key role in enabling sales. One area that highlights this hybrid model is the distribution network, where solar lighting companies often combine efforts with grassroots community mobilizing. Industry stakeholders often draw parallels between the "energy access" market and mobile market, when it comes to market penetration potential for the former. Figure 15.1 illustrates the growth opportunity for "energy access".

According to Navigant Research's report, *Solar Photovoltaic Consumer Products* (Gauntlett 2014), annual revenues for pico solar and solar home systems are expected to grow from approximately $430 million in 2014 to $1.3 billion in 2024. Therefore, not only is the mobile market a suitable analogy for the energy access market – but it is one of the top *enablers* for energy access. The connection between the two is discussed in the following sections in more depth.

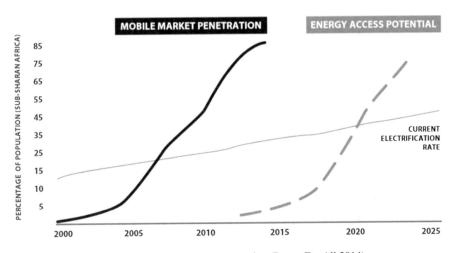

**Fig. 15.1** Sub-Saharan Africa mobile market penetration (Power For All 2014)

## 15.5   A Majority of the Off Grid Population Is Covered by Mobile Networks

As of mid-2013, the GSMA Mobile Enabled Community Services program estimated the global energy addressable markets, e.g. the number of people covered by mobile networks without access to electricity, at more than 643 million people, representing up to 53 % of the global off grid population (Nique 2013). Out of this total, more than 476 million people live in rural areas (Fig. 15.2).

The largest addressable market is Sub Saharan Africa (359 million people) where the reach of electricity networks remains limited (32 % of the population (IEA 2012)) but where mobile networks cover more than 74 % of the population. In East Africa, Kenya, Tanzania and Uganda accounts for more than 82 million people who could benefit from the access to mobile-enabled energy services.

## 15.6   Five Mobile Channels to Enhance Access to Energy

Based on the current footprint and maturity of the mobile industry, five mobile channels can support access to energy solutions (Table 15.2):

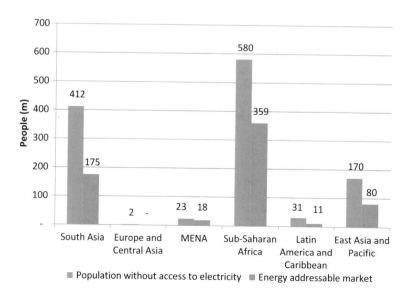

**Fig. 15.2**   The largest addressable market is Sub Saharan Africa

**Table 15.2** Five mobile channels impact summary

| Channels | Impact |
| --- | --- |
| Mobile tower infrastructure | Increase sustainability of decentralized micro-grids by providing power for consumptive and productive use |
| MNO distribution network | Support last mile delivery services for off grid products (e.g. home solar systems) |
| | Improve customer awareness and trust in emerging energy solutions by co-branding products |
| Machine to machine connectivity | Improve maintenance through remote monitoring |
| | Enable Pay As You Go functionality, improving energy solutions affordability |
| | Improved user centric design thanks to consumer usage patterns collection |
| Mobile financial services | Increasing system affordability through Pay As You Go solutions |
| | Improving payment efficiency |
| | Enabling private energy connection finance |
| | Proposing smart tariffs based on customers energy usage or time of usage |
| Mobile services (SMS, USSD, Applications) | Improve utility agents business capability through mobile tools usage (e.g. customer relationship management) |
| | Enable the collection of crowd-sourced information directly through customers |
| | Improve supply chain management through mobile platforms |

## 15.7 Emerging Mobile-Based Energy Business Model—The Pay as You Go Example

While underserved populations spend an important proportion of their income on hazardous energy solutions (up to 30 % of their yearly income mainly on fossil fuels, such as kerosene (Hammond et al. 2007), part of the same population cannot afford to buy clean energy solutions due to their high upfront costs. There are however evidence supporting the hypothesis that offering payment plans to customers increases the rate of adoption of solar products as they become more affordable.

The Pay As You Go model, developed on the brink of mobile financial services deployment, are now allowing entrepreneurs to offer home solar systems under a micro-financed or a "solar as a service" scheme, enabling low income customers to pay for energy directly via their mobile phones:

- Lease to own (micro-financed): An Energy Service Company (ESCo) offers a micro-loan solution to their customers to afford a home solar system; customers first have to make a down-payment to have the home solar system installed at their house and then repay for the full price of the unit part of their energy consumption via daily, weekly or monthly installments. Once they repay for the full cost of the product, they own the home solar system and can use it freely (if the unit is GSM enabled, the unit internal switch will be permanently

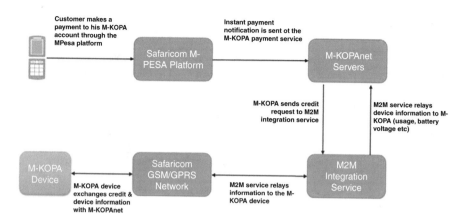

**Fig. 15.3**  M-KOPA pay as you go model

unlocked without any agent intervention). Products usually come with a warranty of 1–3 years according to contract terms.

- Solar as a Service: an Energy Service Company provides a service to its customers while the home solar system remains the property of the service provider. An installation fee is charged to new customers and the service is then provided on a prepaid basis (amount according to the solution capacity). Energy prices are usually lower than in the "lease to own" model as the solar system and other products provided by the ESCo (lights, TV, fridge) remain its property. Full maintenance is also ensured under the service agreement with the end user (Fig. 15.3).

Adding a GSM component to an energy system is the most seamless solution for PAYG, as remote monitoring and credit update on the unit meter can be done over the air, without an agent or a user intervention. Service providers receive real time information under a SMS format about the unit operations (power consumption, battery charge/discharge), customers' payments (frequency of payments, credit) and any maintenance/theft issues. ESCos are also building extensive databases on unit operations, which are then analyzed to offer better products and services from a user centric perspective. M-KOPA, providing home solar systems under a Pay As You Go model in Kenya, has been to date the most successful Pay As You Go provider, reaching 50,000 units sold as of January 2014. The deep integration of mobile technology in their business model, M2M connectivity coupled to mobile payments, added to their distribution partnership with leading mobile operator Safaricom, has enabled this emerging ESCo to provide reliable products and stable services.

Most of the PAYG solutions are currently available in East Africa with poor access to electricity, good mobile penetration and increasing mobile money services traction, this region has become the cradle for mobile-based energy PAYG solution deployments. However, the opportunity to deploy and scale PAYG is real and important in most of the global off grid regions, provided the right ingredients are present: quality energy products, mobile payments capability (using mobile money but also mobile airtime), working capital available to energy entrepreneurs financing the risk of customers default and an efficient distribution network (Table 15.3).

**Table 15.3** Example of energy Pay As You Go providers (as of the end of 2013)

| Companies | Country operations | Service model | Payment type |
|---|---|---|---|
| M-KOPA | Kenya, Uganda | Lease to own | Mobile Money – flexible fee |
| Mobisol | Tanzania, Rwanda | Lease to own | Mobile Money – fixed fee (rates starting at US$12 per month) |
| Off Grid Electrics (OGE) | Tanzania, Ghana | Solar as a service | Mobile Money – flexible fee |
| Angaza design | Kenya, Tanzania | Lease to own | Scratch cards or mobile payments – flexible fee |
| Econet solar | Zimbabwe, Lesotho, Burundi | Solar as a service | Airtime Billing – fixed fee (~US$0.25 per day) |
| Nova Lumos | Guinea, Nigeria | Solar as a service | Airtime Billing |
| Simpa networks | India | Lease to own | Scratch cards |

## 15.8    Challenges of Such Mobile Enabled Solutions

- Financing—while pay as you go solutions improve energy affordability, the financing burden falls on Energy Service Companies providing such products or services. Entrepreneurs have to wait for the duration of the payback periods to recoup their sales revenue, a period that can be as long as 36 months for some providers. This puts an important pressure on cash flow availability, especially for customers with high default risks. Without debt financing tools and working capital, companies might be unable to expand or ensure efficient after sales services.
- Availability of mobile financial services—even though mobile money services are increasingly gaining traction across markets, their growth and how they can be leveraged by energy entrepreneurs will vary on a market basis. The convenience of mobile payments should not be made at the expense of higher fees charged to consumers each time they pay via their mobile phone. In complement to mobile money services, airtime billing represent another interesting opportunity for customer energy payments (Gauntlett 2014).

## 15.9    Discussion

The success of mobile telecommunications in emerging markets is enhancing the opportunity for energy practitioners to develop innovative access models tailored to off grid and underserved communities' ability to pay. According to the GSMA, five mobile channels, based on the reach and impact of the mobile infrastructure, technologies and services, appears key to support such innovation. The development of Pay As You Go (PAYG) solutions under a micro-loan or solar as a service model, where units can be remotely monitored through machine to machine connectivity over mobile networks and where customers can make payments directly via their mobile phones, can act as a paradigm in displacing hazardous fossil fuels for the off grid households with access to mobile. As mobile markets mature and mobile financial services get more traction, most PAYG pilots could move in a commercial phase in 2014, with new entrants in different parts of Africa, Asia and Latin America. Still at a nascent stage of development, technology and business models innovation should lead to the emergence of more synergetic models, coupling mobile, energy but also water, some of the key pillars to socio-economic empowerment.

## References

Gauntlett D (2014) Solar PV consumer products. https://www.navigantresearch.com/research/solar-photovoltaic-consumer-products

Hammond A, Kramer WJ, Tran J, Katz R, Walker C (2007) The next 4 billion: market size and business strategy at the base of the pyramid. World Resources Institute, Washington, DC

IDCL (2014) Infrastructure development company limited. http://idcol.org/home/solar

IEA (2010a) Energy poverty: how to make modern energy access uni-versal? International Energy Agency, United Nations Development Programme, United Nations Industrial Development Organization, Paris

IEA (2010b) World Energy Outlook 2010. International Energy Agency, Paris

IEA (2012) Annual report. International Energy Agency, Paris

IFC (2012) Lighting Africa market trends report 2012. The World Bank, Washington, DC

Lam NL, Smith KR, Gauthier A, Bates MN (2012) Kerosene: a review of household uses and their hazards in low- and middle-income countries. J Toxicol Environ Health B Crit Rev 15(6):396–432. doi:10.1080/10937404.2012.710134

Nique M (2013) Sizing the opportunity of mobile to support energy and water access. GSMA, India

Power For All (2014) The energy access imperative. Power For All, New York

World Bank (2012) Lighting Asia: solar off-grid lighting. International Finance Corporation. The World Bank, Washington, DC

# Chapter 16
# Mobilizing Commercial Investment for Social Good: The Social Success Note

**Lorenzo Bernasconi Kohn and Saskia Bruysten**

**Abstract** In this chapter, we present a new pay-for-success investment model, which we call the "Social Success Note". This approach may offer an innovative solution to crowd-in commercial investment at scale to organizations that seek to achieve both financial sustainability and measurable positive social and environmental outcomes. While initially designed for investment into social enterprises, this pay-for-success financing solution has application wherever there is the potential to incentivize greater positive social and environmental impact through an outcome payment linked to investment.

In the Social Success Note model, a private investor agrees to make capital available to a social enterprise at a below-market rate. The investee has to repay the investment or loan, but if it achieves a predetermined social outcome, a philanthropic outcome payer provides the investor an additional "impact payment" corresponding to a market-rate return.

The Social Success Note is being developed and piloted by Yunus Social Business, a global incubator of social businesses, and The Rockefeller Foundation as part of its Zero Gap portfolio, focused on developing innovative financing solutions to address the world's most critical challenges.

**Keywords** Social enterprises • Innovative finance • Outcomes based financing • Institutional and commercial capital

## 16.1   The Financing Challenge of Social Enterprises

The Social Success Note (SSN) was initially developed as a way of overcoming some of the key challenges to financing social enterprises. Social enterprises are revenue generating businesses which – in addition to seeking financial

L.B. Kohn
The Rockefeller Foundation, New York, NY, USA

S. Bruysten (✉)
Yunus Social Business, Global Initiatives, Co-Founder and CEO, Frankfurt, Germany
e-mail: saskia.bruysten@yunussb.com

© Springer International Publishing Switzerland 2016
E.A. Thomas (ed.), *Broken Pumps and Promises*,
DOI 10.1007/978-3-319-28643-3_16

sustainability – seek to achieve specific social or environmental outcomes. While not necessarily new, social entrepreneurship has exploded over recent years driven by the promise of applying business and management skills to achieving positive social outcomes through a financially self-sustaining model. Around the world, social entrepreneurs are tackling some of humanity's most entrenched challenges from lack of access to energy in Sub-Saharan Africa, to the provision of education to girls in South Asia and access to affordable credit in underserved communities in the United States.

Alongside the rise to prominence of social enterprises, there has been a meteoric rise in investment into the field. According to the latest estimates, there is an estimated $60B invested globally by impact investors into ventures that generate social and environmental impact alongside a financial return, a 16 % increase over the previous year (Global Impact Investing Network 2015). While these headline numbers are impressive, this total represents less than 0.02 % of the estimated $210 trillion in global financial markets (Roxburgh et al. 2011).

According to investors, the key constraint to the growth of the impact investing market is a "lack of appropriate capital across the risk/return spectrum" (Global Impact Investing Network 2015). In other words, there is a disconnect between investors' needs and the risk-adjusted returns that social enterprises can offer. The SSN is designed to address this key challenge.

When the movement around impact investing first emerged, there was an assumption that the traditional impact investors would be willing to forgo financial return for greater social impact. However, evidence suggests that this is not the case. According to the 2015 edition of the GIIN and J.P. Morgan Annual Impact Investor Survey, over 80 % of surveyed impact investors seek "market rate returns" or returns that are "closer to market returns than to capital preservation" (Saltuk et al. 2015).

However, the challenge is that succeeding in any business enterprise is challenging and that succeeding in one with a social mission is doubly so. While there are exceptions to the rule, social enterprises focused on addressing tough and entrenched challenges work in situations where there are significant market failures that result in lower returns. For example, selling rooftop solar providing affordable and clean energy to individual households in a developed market such as the United States is much easier than doing the same in an underserved market such as, say, Uganda. In the United States, there is a whole set of enabling conditions such as access to credit for consumers, a well-developed transportation infrastructure, government tax incentives etc. that do not exist in Uganda. As a result, most social enterprises face an inverse relationship between profit margins and social impact – the more intractable and difficult the problem they seek to solve, the more difficult it is for them to achieve market rate returns. This challenge is generalizable to businesses beyond social enterprises where there is an opportunity to achieve social or environmental impact which needs to be balanced with financial return.

This disconnect between the need for financial-return on one side, and the maximization of impact on the other, gives rise to two suboptimal potential scenarios. The first scenario is that the field of social entrepreneurship remains subscale and largely funded by philanthropic dollars. Philanthropy, even when combined with government, does not have anywhere near the resources to address the world's most pressing challenges. The United Nations, for example, estimates that it will cost an estimated $3.9 trillion a year to achieve the Sustainable Development Goals (SDGs) over the next 15 years in developing countries alone. However, current levels of public and private funding only cover $1.4 trillion, leaving an annual shortfall of $2.5 trillion (World Investment Report – Investing in the SDGs: an action plan 2014). By contrast, the combined assets of global foundations and Development Finance Institutions (DFIs) comes to approximately $2.5 trillion of which only a fraction is lent or donated on an annual basis.

The second scenario is that social enterprises increasingly fall prey to "mission drift" whereby social impact is sacrificed for financial return. Mission drift is a widespread challenge for growing social enterprises as they seek additional capital for growth. The short history of microfinance illustrates some of the perilous pitfalls of compromising social mission for impact (SKS Under Spotlight in Suicides 2012).

The SSN looks to develop a third scenario: one where the needs of investors are met while ensuring mission-focus of social entrepreneurs looking to resolve the world's biggest challenges.

## 16.2  The Social Success Note: Crowding in Commercial Capital for Impact

The SSN harnesses the power of pay-for-success contracts which lies at the heart of a wave of new financing instruments to foster positive social or environmental impact highlighted most prominently by Social Impact Bonds and Development Impact Bonds.

Pay-for-success is not, as such, new. Performance contracting for infrastructure or energy efficiency, for example, is commonplace and has been around for years (Fostering Investment in Infrastructure 2015). The innovation with these pay-for-success financing models is that they crowd-in *investment capital* to where it would otherwise not flow through the pricing and transfer of risk. An overview of how the SSN works helps illustrate this point (Fig. 16.1).

Let us take an example of a hypothetical social entrepreneur, Amira, who is focused on selling solar panels for homes across sub-Saharan Africa with a focus on underserved, poorer consumers. Amira would like to expand her business to new markets, but can't attract the right kind of investment. Her business like many others, is too small for domestic commercial banks and too big for microfinance institutions while a traditional private equity firm focused primarily on financial returns would compromise the social mission of her business. The SSN offers a solution to Amira's problem. Like other pay-for-performance financing structures, it involves three parties: a business, an investor, and a donor. In the SSN agreement, a commercial investor agrees to provide Amira with a concessionary loan of, say, $1 M at zero percent. Amira's business must pay back this loan. However, if Amira

**Fig. 16.1** An illustration of the Social Success Note in practice

achieves a pre-determined social outcome of, for example, reaching 10,000 new household customers (verified by a third-party evaluator), the donor pays the investor an "outcome payment" that amounts to a competitive market-rate return. As this example highlights, through the pricing of social outcomes (i.e., the value to the donor of expanding the reach of solar to difficult-to-serve markets) and the transferring of the risk to the private sector (i.e., reflected in the interest premium), the SSN is able to crowd-in investment capital to achieve social and environmental outcomes that would otherwise not be commercially financed.

The resulting financing structure of the SSN has a number of attractive features. First, the social entrepreneur receives low-cost financing to pursue both his/her business and social goals, without the need to compromise one of the two. Indeed, the SSN actually helps to reinforce the mission, since the pay-out is conditional on demonstrated social (or environmental) impact. The investor, in turn, achieves an attractive market return by investing in social outcomes that are likely to be less correlated than other commercial investments with the market thus helping him/her diversify risk. Finally, the donor achieves large social outcomes for a fraction of the normal cost, and at no risk.

As illustrated in Fig. 16.2, the SSN shares many similarities with Social Impact Bonds, not least the fact that it is focused on funding *results* rather than inputs. However there are four key differences which are important to highlight in order to understand the distinctive scope as well as advantages/disadvantages of the SSN model:

- **The SSN applies beyond cost-saving.** SIBs capitalize on an important and evident but too often ignored insight which is that upfront investments in preventative measures can generate enormous long term social and economic benefits for individuals and communities. SIBs capitalize on this insight but go a step further in addressing a missing working capital need that NGOs and other service providers need. In the case of a large capital project such as building a road, developers can access upfront costs independently to do the work. This is not the case for most non-profits (Keohane 2016). As illustrated in Fig. 16.1, by monetizing

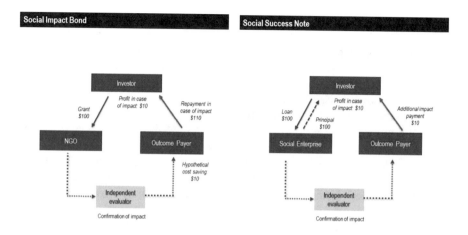

**Fig. 16.2** Comparison of the mechanics of a Social Impact Bond (SIB) and Social Success Note (SSN)

the future value of cost savings, SIBs mobilize new sources of private investment to supply the working capital for the service providers to undertake the interventions. SIBs uniquely bring together this partnership model. However, the SIBs model only works where there are measurable cost savings for committed budget line items. This is one of the reasons SIBs applies most directly to governments which are fiscally constrained and unable to take the operational risk of testing new, potentially cost saving measures. The SSN model, by contrast, may apply to cost-saving measure but not exclusively. For example, in the case of Amira, the solar social entrepreneur described above, the philanthropic outcome payer is paying for specific outcomes (households with access to electricity), not cost-savings.

- **The SSN represents a long-term solution to crowd-in new, return-seeking capital to achieve social and environmental outcomes.** The power of the SSN model is that it offers an elegant way of aligning the incentives of entrepreneurs, investors and donors in a manner which is replicable and scalable over time. It applies wherever there is the potential to incentivize greater positive social and environmental impact through an outcome payment linked to investment. By contrast, the principal value of SIBs is in identifying inefficiencies in the delivery of social services by government (Overholser 2015). As a model for crowding-in private sector investment to achieve greater social outcomes over time, the SIBs model is not very efficient. After all, it relies on government paying a premium to the investor to front-load an investment to an NGO to carry out a preventative, potentially cost-saving intervention. If, however, this intervention proves successful, it would be inefficient for government to continue paying this premium – a better route would be to pay the NGO directly. This is indeed what happened in the case of the first SIB pilot ever, launched in 2010 in Peterborough,

UK, to reduce rates of recidivism. Even though the SIB did not achieve the required reduction in recidivism to trigger immediate payment to investors, the UK government decided to end this SIBs program early and extend similar rehabilitation interventions across the whole of the UK as this was more cost efficient than the prospect of potentially paying-out to investors.

- **The SSN spreads risk to investors facilitating innovation.** Another important difference between the SSN and SIBs is around the risk profile to investors. The SSN investor's risk is split two-ways: the risk around the principal is related to the credit risk of the social enterprise while the impact payment bonus is related to whether or not the social enterprise achieves a predetermined outcome. This is very different to the SIB model where all risk is related to the outcome. As a result, the SSN offers more flexibility to invest in higher-risk and more innovative solutions than the traditional SIB model which relies on interventions with an established evidence base which is often difficult to come by and thus limited their scope.

- **The SSN works with outputs, not just outcomes.** Like SIBs, the SSN model relies on making payments only after certain results have been demonstrated. However, the types of results that are amendable to the SSN model are more flexible than what has traditionally been the case with SIBs. SIBs deal with outcomes rather than outputs; they deal with how people's lives are affected by an intervention rather than how many people have been served. This makes sense in the case of assessing cost-saving interventions because, ultimately, it is these outcomes that drive a reduction in cost. As the SSN model is not dependent on cost-savings, there is more flexibility in terms of the results that serve as triggers for payment. Outputs, such as the number of new household customers reached can serve as the results trigger provided that this serves as a robust proxy for positive outcomes that the donor cares about (e.g. reducing lack of access to energy and fighting climate change). Relatedly, the burden of proof necessary for proving the link between an intervention and an outcome is much more complicated to establish than outputs. For example, in the case of the Peterborough SIB, proving success required rigorous measurement and evaluation and the introduction of a "randomized control" group of prisoners who would not receive services in order to establish the impact of the intervention. The high transaction costs involved with this evaluation process has meant that SIBs remain relatively small, bespoke and time consuming as well as expensive to structure.

Our analysis above has focused on the application of the SSN to social enterprises. However, as suggested above, the mechanism applies wherever there is the potential to incentivize greater positive social and environmental impact through an outcome payment linked to investment. For example, we could imagine an SSN type structure that is application to a Green Bond market in order to incentive issuers such as large multinationals to achieve specific outcomes through the use of proceeds of their capital raising.

However, even if we limit the scope of the SSN to social enterprises, we believe that the mechanism represents a very sizable opportunity. For example, if development agencies channelled as little as two percent of their Official

Development Assistance (ODA) through an SSN structure representing a sum of $3 Billion USD, $12.5 billion in commercial capital could be crowded-in to achieve greater social and environmental outcomes (Assumes a leverage of $4 in commercial capital for every $1 in grant funding and an average market investment return in developing countries of around 8 % p.a. over a 3 year maturity note.)

## 16.3 Operationalizing the Social Success Note Model

The sections above have highlighted the scope of the SSN and its promise as a new financing mechanism to mobilize private sector capital for social good. Although the launch of the first SSN pilot is still under development, early learnings have emerged from our research on the some of the key success factors for operationalizing the SSN model:

1. **Ensure a robust legal setup**
   The legal setup to facilitate the SSN model will vary across jurisdictions and require different structures such as Special Purpose Vehicles (SPV) to facilitate donors to act as outcome payers.
2. **Choose the right proxies for the outcomes you want to achieve**
   As explored above, the SSN works with outputs as well as outcomes to trigger payment. While this offers greater flexibility, it is important to ensure that chosen outputs serve as robust proxies for desired outcomes. For example, based on historical data around use and product life, the installation of a particular solar technology in a region may suffice as an indicator of access to energy to individual households.
3. **Choose to the shortest timeframe possible to keep the financing costs of the SSN down**
   The financial viability of any SSN transaction will depend critically on the timeframe of assessing a results payment. The longer the time-lag for returning capital back to investors, the more expensive the transaction. It is therefore important to choose payment triggers that are as close as possible to when the investment of the SSN transaction occurs.
4. **Use objective, easily verifiable outcome payment triggers**
   SIBs have suffered from crippling transaction costs related to verifying outcomes making them difficult to scale and replicate. The flexibility of the SSN should be leveraged to choose outcome payment triggers that are objective, and easily verifiable facilitated, if possible, by the use of technology to reduce transaction costs as much as possible.
5. **Be watchful of the interest rate environment in which you launch an SSN transaction**
   The leverage potential that donors can benefit using an SSN structure is very sensitive to the interest rate environment in which the SSN transaction is

launched. The higher the "market-rate" interest rate, the larger the impact payment to investors. In general, emerging markets have higher interest rates reflecting, among other factors, higher risks meaning that the leverage potential for donors will be lower in these markets.

# References

Fostering Investment in Infrastructure (2015) OECD report, vol 2015. OECD, Paris

Global Impact Investing Network (2015) GIIN and J.P. Morgan annual impact investor survey. https://thegiin.org/knowledge/publication/eyes-on-the-horizon

Keohane GL (2016) Capital and the common good: how innovative finance is tackling the world's most urgent problems. Colombia University, New York

Overholser G (2015) The payoff of pay-for-success: responses. Stanford Social Innovation Review Fall 2015, Stanford Palo Alto

Roxburgh C, Lund S, Piotrowski J (2011) Mapping global market 2011. McKinsey Global Institute, New York

Saltuk Y, Idrissi AE, Bouri A, Mudaliar A, Schiff H (2015) Eyes on the horizon, the impact investor survey. GIIN & J.P. Morgan, New York

SKS Under Spotlight in Suicides (2012) The Wall Street Journal. http://online.wsj.com/article/SB10001424052970203918304577242602296683134.html

World Investment Report – Investing in the SDGs: an action plan (2014) United Nations conference on trade and development, New York

# Index

© Springer International Publishing Switzerland 2016
E.A. Thomas (ed.), *Broken Pumps and Promises*,
DOI 10.1007/978-3-319-28643-3

Printed in the United States
By Bookmasters